CW00727362

Political Economy of Gender and Development in Africa

Bhabani Shankar Nayak
Editor

Political Economy of Gender and Development in Africa

Mapping Gaps, Conflicts and Representation

Editor
Bhabani Shankar Nayak
University for the Creative Arts
Epsom, UK

ISBN 978-3-031-18828-2 ISBN 978-3-031-18829-9 (eBook)
https://doi.org/10.1007/978-3-031-18829-9

This Palgrave Macmillan imprint is published by the registered company Springer Nature Switzerland AG.
The registered company address is: Gewerbestrasse 11, 6330 Cham, Switzerland

About the Book

This volume on '*Political Economy of Gender and Development in Africa; Mapping Gaps, Conflicts and Representation*' is driven by issues rather than abstract theories on gender and development. The book offers clear discussions of central issues of conflict, representation, accessibility and availability of resources to materialise gender justice and ensure rights of women in different spheres of private and public lives. The book helps us to think and understand different issues and challenges of gender and development in Africa. The aim of this book is to focus on threads connecting with the present needs of women to a possible future emancipation of women from the institutions, structures and processes that perpetuate poverty and accelerate different forms of uneven development. The ideals and practices of both gendered development and patriarchal policy processes as contested domains, where capitalism, patriarchy, class, race and other state power structures are undermining women's empowerment within existing political, economic, social, cultural and religious systems.

Contents

Notes on Contributors

Fathima Azmiya Badurdeen works as a lecturer at the Department of Social Sciences, Technical University of Mombasa, Kenya. Since 2012, she has worked in Kenya as an academic and practitioner in the field of forced migration, human trafficking and countering violent extremism. She has increasingly used ethnographic methods and performance art in research work. Prior to her work in Kenya, she has worked as a researcher and practitioner in Sri Lanka in the field of conflict transformation and peace building. Most of her publications are available online.

Fawzi Dekhil has been a university teacher in Marketing and Management since 1997 and is a full Professor in Marketing since 2020 at the Faculty of Economic Sciences and Management of Tunis. He received the PhD in marketing from the University of Tunis El Manar in 2006. Since 2018, he has been responsible for the Master of Marketing Research. He has several publications, conference communications and book chapters to his credit. His research interests are in the areas of Marketing & Communication, Sponsoring and Sports Marketing, Consumer Behaviour, Advertising and Digital Communication, Service Marketing (tourism, banking), Cultural and Islamic marketing, Marketing Research, Food Product and Food Waste and Kid's Marketing. He was an Advertising Manager of the Coca-Cola company in Tunisia between 1995 and 1997. He is a member of scientific committees and organizing committees in several conferences. He is a member in several scientific, sport and cultural associations and a member in several national commissions. He is also a referee reports for several Scientific Journals. He is a founding member of the "Tunisian

Marketing Association" (ATM) and a member of "ERMA" laboratory. Since 1997, he has been a marketing & advertising consultant. Since 2008, he has been a marketing and communication trainer.

Linda Etale is a Research Associate at the Africa Centre for Evidence at the University of Johannesburg. She is also pursuing her Post-Doctoral Fellowship on gender and agricultural research at the CGIAR GENDER Platform, based in Nairobi, Kenya. She joined the CGIAR GENDER Platform to support the development of a gender research portfolio to help address social equity objectives. She also has vast experience in institutional strengthening and grants management. Her interests are in evidence-based research and implementation of development programs in community development, gender, land rights, food security, environmental integrity and climate change. Her interest is in contributing to evidence-based research that positively impacts the socioeconomic status of agriculture-dependent communities while sustainably addressing gender inequalities. She has a growing passion for communicating science to lay audiences and believes in sharing scientific knowledge with the public. Linda has worked on projects in Africa involving government agencies, non-governmental organizations and research institutions. Etale received her PhD in Geography and Environmental Science from the University of Witwatersrand, South Africa in 2020. She also holds an MSc degree in Education for Sustainability and a BA in International Relations.

Mouna Ben Ghanem is an interdisciplinary researcher. She received her PhD in Marketing from the University of Tunis El Manar, Faculty of Economic Sciences and Management of Tunis, Tunisia since 2017. Her research revolves around advertising creativity, rhetoric, political discourse, digital communication. She has written papers and presented at numerous conferences and events. She is a member of "*ERMA*" laboratory.

Bhabani Shankar Nayak is a political economist and works as Professor of Business Management and Programme Director of Strategic Business and Management at the University for the Creative Arts, UK. His research interests consist of closely interrelated and mutually guiding programmes surrounding political economy of religion, business and capitalism, along with faith and globalisation, and economic policies. He is the author of *Political Economy of Development and Business* (2022), *Modern Corporate Strategies at Work* (2022), *China: The Bankable State* (2021), *Disenchanted India and Beyond: Musings on the Lockdown Alternatives* (2020), *Hindu*

Fundamentalism and the Spirit of Global Capitalism in India (2018) and *Nationalising Crisis: The Political Economy of Public Policy in India* (2007).

Douglas Onwumah is an economist who works as a consultant in private sector in the United States of America. He received his MSc in Economic Development from the Adam Smith Business School, University of Glasgow, Scotland, UK. His primary research interests are in the field of economic development.

Madelin O'Toole is an Economist with the Federal Research Division at the Library of Congress in Washington DC. She holds a Master of Science in Financial Economics from the Adam Smith Business School at the University of Glasgow and a Bachelor's of Arts in Economics from Christopher Newport University. Her research interests include development, financial, gender and conflict economics.

Mulala Simatele is based at the Global Change Institute (GCI) at the University of the Witwatersrand. He is an Environmental Scientist by training and specialized in Geographies of the Environment and Sustainability. Simatele worked for the University of St. Andrews in Scotland where he was part of the team which established the St Andrews Sustainability Institute. His main areas of research interest revolve around community-based natural resource management, education for sustainability, water and sustainability, climate change adaptation, environmental justice, environmental impact assessments, marine resource management, and disaster risk management. In addition to academic engagement, Simatele is one of three technical advisors to the International Development Research Centre (IDRC) and the Australian Center for International Agricultural Research (ACIAR). He is also a board member of Hanell International and has diverse experience working with policy-related institutions notably on environmental management, climate change adaptation and environmental sustainability. Notable among these institutions include The World Bank, where he served as the Chief Environmental Officer in the Climate Change Unit advising different Country Directors in Africa and Asia on climate change and environmental issues, and The Scottish Environmental Think Tank, where he served in the position of Deputy Director. He has also served as an environmental consultant for different governments: the government of Bolivia, Botswana, Ghana, Jamaica, Kenya, Malawi, Namibia, Tanzania, Sweden, Zambia and Zimbabwe. He has also worked with various NGOs on environmental and climate change issues including community development. He holds a DPhil in Environmental Management and Sustainable Development from the University of Sussex in the UK. He has co-authored over 60 publications covering climate change adaptation, solid waste management, environmental justice, natural resource management, among others.

Geeta Sinha works as Associate Professor in Public Policy. She is trained in International Development, Gender and Social Sciences. She received her Doctoral Fellowship from the Oxford Brookes University, UK, where she is completing her PhD in Development Economics from the Oxford Brookes Business School. She received the International Ford Foundation Fellowship and obtained her MA in Gender and Development from the Institute of Development Studies (IDS), University of Sussex, UK. She holds a Postgraduate Diploma in Rural Development with specialisation in Project Management from Xavier Institute of Social Service (XISS). Prior to joining the O.P. Jindal Global (Institution of Eminence Deemed To Be University), she worked with the Regent's University London as Visiting Lecturer in the School of Humanities and Social Sciences. Besides these, she has extensive research and consultancy experience with the MART, PRAXIS, Development Alternatives, Aide et Action and NGOs like Jan Jagran Sansthan & JPMVK. She worked extensively in livelihoods and enterprise development, skills development, agribusiness, value chain development, women empowerment, poverty alleviation programs, and other research impact studies in different parts of India, Bangladesh and Nepal. She is completing her PhD thesis on "mining led industrialisation and gender-based violence amongst indigenous communities in Odisha, India". Her research unravels gendered impact of mining-led industrialisation on women and on indigenous communities through Eco socialist and feminist lenses. She reviews articles for international journals such as *International Journal for Finance and Economics, Journal of Asian and African Studies, Journal of Global Entrepreneurship Research, Financial Innovation and World Journal of Entrepreneurship, Management and Sustainable Development.* She has published research papers in peer reviewed journals and edited books.

List of Figures

List of Tables

Introduction: Political Economy of Gender and Development in Africa

This volume on '*Political Economy of Gender and Development in Africa; Mapping Gaps, Conflicts and Representation*' is driven by issues rather than abstract theories on gender and development. The book offers clear discussions of central issues of conflict, representation, accessibility and availability of resources to materialise gender justice and ensure rights of women in different spheres of private and public lives. The book helps us to think and understand different issues and challenges of gender and development in Africa. The aim of this book is to focus on threads connecting with the present needs of women to a possible future emancipation of women from the institutions, structures and processes that perpetuate poverty and accelerate different forms of uneven development. The ideals and practices of both gendered development and patriarchal policy processes as contested domains, where capitalism, patriarchy, class, race and other state power structures are undermining women's empowerment within existing political, economic, social, cultural and religious systems.

Common sense and stereotypes on gender and development in Africa are dominant ideologies that underpin different theories of development of African women. The book questions such narratives. It is a critique of political economy of gender and development in Africa and re-imagining the existing theoretical outlooks from an issue-based perspective moving beyond essentialist and functionalist narratives. It intends to move within the terrain of dialectical approach of overcoming different theoretical dominance over 'gender and development' within bourgeois thought and transforming them at the same time. The issues outlined in this book are interconnected to understand and articulate a radical and pluriversal agenda of transforming 'women's empowerment to women's emancipation'. It is within this spirit; the book makes significant interventions in different theoretical debates on gender and development within the African context.

African women need African consciousness, ideas, ideologies, structures and agencies for the creative destruction of patriarchy and capitalism in Africa. However, the book rejects the categorisation of 'gender' and 'politics of women' as identities that subconsciously, prevent greater social and political solidarity for emancipatory struggles. Marginalisation within capitalism and patriarchy is a process which established its legal, political, cultural, religious and economic institutions to ensure unequal ownership, access and right to resources for an egalitarian and dignified life as women. Such gender-based structural barriers are not confined to Africa, it is an universal story of proletarianisation and marginalisation of women within the processes of development. The different forms of marginalisation of women within different processes of development continue to create material circumstances that are common to women across the world. Therefore, organic solidarity among women transcends territorial boundaries of Africa. The issues of livelihood, property rights, food security, representation, accessibility to resources, environment and development empowerment are not only the issues of African women. These are issues of all women in the world.

In the context of Africa, patriarchy, racial and colonial capitalism have not only reinforced gender inequalities but also accelerated uneven development where women were alienated from the processes of development. Such a disempowering process has weakened the citizenship rights of women in Africa. Therefore, it is important to locate issues and their interconnectedness in totality to see through the racial and patriarchal capitalist social, political, economic and cultural relations established by European colonialism and its modernisation project and development processes operating in Africa with different institutional mechanisms.

Linda Etale and Mulala Simatele's chapter argues for women's right to land for meaningful growth of agricultural economy in Kenya. The chapter highlights legislative contradictions and cultural impediments within legislative frameworks for the women's right to land. Mouna Ben Ghanem and Fawzi Dekhil's chapter identifies the portrayal of men and women in television advertisements during pre- and post-Tunisian revolution. In Chaps. 3, 4 and 5, Madelin O'Toole and Bhabani Nayak argue for availability and accessibility of inclusive finance for women to move forward in a progressive path of inclusive and sustainable economic growth in Rwanda. In Chap. 6, Douglas Onwumah and Bhabani Nayak identify the negative and positive role of aid for economic growth in Nigeria. Geeta Sinha and Bhabani Nayak argue for mainstreaming of gender and education with the help of sector-wide approaches for the empowerment of women in Ghana. The final chapter explains sexual and gender-based violence (SGBV) among Somali refugee women and their inner resilience to survive in the Dadaab refugee complex in Kenya.

Epsom, UK Bhabani Shankar Nayak

CHAPTER 1

Mapping Contradictions Within the Legal Frameworks and Cultural Norms on women's Right to Land and Agriculture in Western Kenya

Linda Etale and Mulala Simatele

Introduction

The importance of land as a resource to the Global South agriculture-dependent rural populations cannot be overlooked in the development agenda, particularly for women's empowerment. Secure land rights for rural women are crucial in ensuring food and economic security of their households (Galiè et al., 2015; FAO, 2020), as there is likelihood of

L. Etale (✉)
Africa Centre for Evidence, University of Johannesburg, Johannesburg, South Africa

M. Simatele
Global Change Institute (GCI), University of the Witwatersrand, Johannesburg, South Africa
e-mail: mulala.simatele@wits.ac.za

B. S. Nayak (ed.), *Political Economy of Gender and Development in Africa*, https://doi.org/10.1007/978-3-031-18829-9_1

higher levels of agricultural investment and productivity, increased income, and household food security (Galiè et al., 2015; Mishra & Sam, 2016; FAO, 2020). FAO (2020) posits that secure land rights provide women with greater bargaining power especially with their men counterparts, and protect them from falling into poverty.

Women in Kenya play a critical role in Kenya's agricultural ecosystem, making about 59% of the agricultural labor force (ILO, 2019). However, their access is not commensurate with their effort in the sector and along its value chains. Access to land for rural women in Kenya is usually pegged on cultural norms, which prescribe women's access to land through their male relations including marriage; a situation that most often does not guarantee secure access particularly for widows and divorced women. Citing Anukriti (2014), Mishra and Sam (2016) argue that this kind of reality results in discrimination which has throughout history been perpetuated by societal views of women as economically less productive and of lesser value to parents as regards long-term asset accumulation. Kenya has put in place progressive legal frameworks that champion the rights of women to land, however these have not guaranteed secure tenure of rights. A conundrum exists due to a seemingly clashing of frameworks, that is cultural norms versus legal frameworks, which continue to exist and influence decision-making on women's land rights in Kenya. This has not been well analyzed, particularly in the context of Western Kenya.

The objective of this paper is to bring to the fore the contradictory existence of legislative frameworks and cultural norms upon which land administration in Kenya is anchored, and how this reality affects rural agriculture-dependent women in Western Kenya. By doing so, we aim to explore ways in which this complexity can be unraveled, and how women's important role in rural agriculture can be protected through secure land rights. We acknowledge the importance of cultural norms but also make a case for the need to expedite the implementation of legislative frameworks where women's rights continue to be disregarded on the basis of cultural expectations. The premise of this is anchored in the supreme law—the Constitution of Kenya, 2010—which voids any law, including customary law that is inconsistent with this Constitution itself.

Women's Participation in Agricultural Production in Kenya

The rural population in sub-Saharan Africa (SSA), including Kenya, has agriculture as their main occupation, with a majority being smallholder farmers (Davis et al., 2017). Agriculture is the backbone of Kenya's economy contributing approximately 34.2% of the country's gross domestic product (GDP) and employs over 60% of the country's population of which 70% live in rural areas (Kihiu & Munene, 2021). In view of this, the sector has direct implications on food security, livelihood development, and employment; all these directly impact poverty levels and quality of life (Onyalo, 2019; Kansiime et al., 2021).

Palacios-Lopez et al. (2017) argues that women in SSA produce a considerably large percentage of food for both household consumption as well as cash crops from small parcels of land. Women play an integral role in Kenya's agricultural production, working as both unpaid workers in family farms and as unpaid workers in other farms and agricultural enterprises (Onyalo, 2019; Diiro et al., 2018). Women in rural agriculture manage approximately 40% of the smallholder plots (Kihiu & Munene, 2021) and their roles are notable in all stages of food and livestock production. They are involved in land preparation, weeding, harvesting, storage, seed harvesting, and marketing of produce for both subsistence and cash crops. In animal production, women are mostly known for poultry keeping and livestock keeping, particularly for small ruminants.

Despite their critical role in the agriculture sector, rural women face a myriad of challenges fostered by persistent gender inequalities in respect to access and control over productive resources (Onyalo, 2019). Additionally, a survey by the World Bank in Kenya identified impeded access to water, energy and affordable finance, inputs and transport to markets as other areas of burden for women (World Bank, 2011). They also face challenges in access to information and formal extension, as Kansiime et al. (2021) report that women's access is generally low at 33%, noting that men were more likely to access such services than women. Access, control, and ownership of land as a capital of production are gendered, and mediated by patriarchal customary laws (Kameri-Mbote & Muriungi, 2018). As explained by Munala (1995) and Karanja (1991), land in Kenya is mainly passed down through generations in patrilineal succession in a patrilocal manner through permanent members of the family, who are men in this case.

Literature Review: Legal Frameworks, Cultural Norms, and Women's Land Rights in Kenya

Women in Kenya, like those in other SSA countries, continue to contend with the challenge of insecure access to land for the purposes of livelihood development and securing adequate nutrition for their dependents (Musangi, 2017). Recent estimates show that only 1% of women in Kenya are sole owners of land (titled land) and only 5% are registered owners of land with their husbands (Kameri-Mbote & Muriungi, 2018). FIDA-Kenya estimates that even though women's ownership of land through titling is very low compared to men, they provide 89% of labor in subsistence farming and 70% of labor in cash crop farming. This reality stems from cultural norms that have long dictated that women cannot own or control land; they can only have customary rights through consent from their male relations to access and cultivate land (IWHRC, 2008). From a Neo-Marxist perspective in smallholder contexts as explained by Meillassoux (1981), in a patriarchal society such as Kenya, women are exploited for labor by men who own and control capital of production. Even in contexts where they are engaged externally as casual laborers, women's subordination makes them susceptible to exploitation of their labor and which results in them not fully remunerated (Oya, 2011). Bikketi et al. (2016) posit that women's lower social status that dictates their access to capital of production can be traced to a persisting socialization process that ensures daughters are brought up to be docile to their husbands. This kind of reality stems from patriarchy that dictates intra-household power structure where women are subservient to the dictates of their husbands or male relatives (Ibid). A study in Western Kenya confirms that power relations that give men the power to control access and use of land within the marital context become more complex in polygamous settings; wives in such marriages have their power reduced unless each co-wife has been allocated their own plot (Ibid). In cases where women purchase their own parcels of land, their husbands demand sole ownership (including through titling) and usually control use (Bikketi et al., 2016).

Kenya has made notable strides in terms of ensuring legal frameworks secure women's land rights (Kameri-Mbote & Muriungi, 2018). Even with such progress there are challenges that exist even with expectations set in regional and international law (such as the Protocol to the African Charter on Human and People's Rights on the Rights of Women-the

Maputo Protocol) that is emphatic on states to ensure their laws guarantee equitable sharing of joint property derived from marriage upon its dissolution.[1]

The Constitution of Kenya (COK 2010) which is the supreme law is very clear by calling for the elimination of any form of discrimination based on gender, customs, and practices—this includes land (Government of Kenya, 2010). The COK, recognizes the right of women to equal treatment under the law and it emphasizes that it supersedes cultural norms, particularly where anyone is discriminated based on their gender. Under Article 27 of the COK, women's rights to land are legally equal to those of men, even though the reality is that there is a significant gap between men's and women's right to land (LANDESA, 2014). The significant gap is mainly caused by cultural norms that have kept women from accessing their rights to land due to patriarchal rules that ensure land ownership and control is the preserve of men.

It is very clear that the Constitution of Kenya (COK), and other legislative frameworks do not limit women on the land that they can own, but on the other hand customary rules do. The Matrimonial Property Act, 2013, directs that couples in a marriage have equal rights to property including land which is in tandem with provisions of the COK, (Government of Kenya, 2013; LANDESA, 2014). This Act prescribes that application of customary law should not contradict the principles of the COK 2010. There are limitations to the Matrimonial Property Act, 2013, a situation that leaves women fighting to secure their parcels of land upon death or divorce from a husband because the Act only applies where the spouses acquired the land together (Mbugua, 2018). Women fighting to secure ancestral land acquired through marriage cannot benefit from the protection of this piece of legislation. To secure matrimonial property, Section 79 (3) of the Land Act 2012 (Government of Kenya, 2012) provides that such property cannot be disposed of by a registered owner short of the consent or knowledge of the other spouse. Similarly, the Land Act, 2012, directs that before a matrimonial parcel of land is sold, it is mandatory for spousal consent to be obtained (Government of Kenya, 2012). In the case of death of a father or husband, the Law of Succession Act, Cap. 160, safeguards the inheritance rights of women to land.

The Land Registration Act, 2012, protects women's land rights through allowing joint tenancy and securing use by women, which did not exist before the promulgation of the 2010 Constitution. The Act also directs that disposition of land that is co-owned between spouses be done through consent. In the context of community land the rights of women are secured. Section 14(4) c (i and ii) of the Community Land Act, 2016, stipulates that regarding registered communal land, there shall be equal treatment of application for women and men and there shall be no discrimination based on gender or marital status (Kameri-Mbote & Muriungi, 2018). Anticipating disputes and challenges on land and property between men and women, for the first time in Kenya's history, an Environment and Land Court was established as per the directive in the COK 2010 through the Environment and Land Court Act, 2011. This court is mandated with determining disputes relating to land.

The Predicament of Legal Versus Cultural Norms in Kenya

Notwithstanding the clear relationship between non-discriminatory secure land rights and sustainable agriculture and food security, women in Kenya have barely enjoyed these rights, given the gendered nature of access, control, and ownership, where such access and use are mediated by patriarchal customary laws (Kameri-Mbote & Muriungi, 2018). The reality is that all matters of land are still addressed through dual lenses, that is, customary laws and contemporary law.

Burke and Jayne (2014) underline that all over SSA, women are generally disproportionately marginalized in the distribution of land. This holds true in Kenya and particularly in Western Kenya, where patriarchal traditions still dictate that sons inherit and control land from their fathers to continue the lineage. This is premised on the notion that sons remain within their families, and they would protect such property, while daughters are expected to get married and leave. FIDA posits that perpetuation of this norm gives precedence to male relatives, and terms this as a guise of male "protection" which actually strips women of their property. The human rights organization further states that such practice serves as a way of asserting control over women's autonomy and labor while enriching their "protectors" (FIDA). When a woman gets married and leaves her father's land, her marital situation is not any better because, upon the death of her husband, she is often disinherited of her husband's land and property; such women rarely are aware of their rights, and in case they are, they are unlikely to have the resources and the know-how to pursue justice especially through legal means (FIDA; Burke & Jayne, 2014). Additionally widows are pressured to get married to brothers of their deceased husbands and any land and property in their possession may be reallocated to her new husband or other family members (Yamano et al., 2009). Notably, even when relevant laws to sort disputes over land are available, there are barriers to understanding as these are written in technical language which is not easily comprehensible by rural women.

The complex mix of legal versus cultural norms of dealing with gendered land rights in Kenya often leaves women more vulnerable to unfair practices than actually securing their rights. This is mainly due to the reality that in rural contexts cultural norms still dictate decision-making despite the government's enacting relevant laws which supersede the cultural dictates. This challenge often stands in the way of realizing gender equality as far as land rights are concerned.

Feminism Challenging Patriarchy

Feminism as a development paradigm challenges patriarchal norms by espousing the equality between men and women in all aspects of life including access and control of capital of production. It is important to note that feminism does not foster the exclusion of men or only furthering women's causes (Ibid). For this study, we did not focus on specific branches of feminism (liberal, radical, socialist, and womanism) but rather chose to use the paradigm in general. At the core of feminism is creating change that removes inequalities by doing away with power imbalances, in this case brought about by patriarchal norms. By doing so it also starts to look at standpoints and experiences of women in different as well as specific contexts, while striving to understand women's oppression with the purpose of ending it (Ardovini-brooker, 2002).

Context of the Study

Kakamega County is predominantly a crop farming economy with 61% of its population engaged in the various agricultural activities, fewer than 255,483 ha of land (MoALF, 2017). We conducted this study in four wards in Matungu and Mumias East sub-counties of Kakamega County in the western region of Kenya where agriculture is one of the mainstays, with an average land holding of 1.5 acres. The four wards are *Malaha-Isongo-Makunga* and *East Wanga* both in Mumias East Sub-County and *Namamali* and *Khalaba* in Matungu Sub-County. The study sites have been mainly dedicated to sugarcane farming as the main cash crop targeting local millers; however, subsistence farming is also practiced within the same land holdings. The study area receives an annual rainfall that ranges between 2214.1 mm and 1280.1 mm per year, temperatures range between 18°C and 29°C, average humidity is 67%, and altitude ranges from 1240 to 2000 meters above sea level (The County Government of Kakamega, 2013).

The dominant research paradigm was qualitative even though quantitative techniques were used to gather specific data, a decision that was influenced by Westmarland painting out the dangers of relying on one paradigm over the other. She argues that linking feminist research with qualitative methods simply reinforces traditional dichotomies that may not be in the best interests of feminist research. A framework analysis as well as an inductive approach were used to make sense of the data. Feminism allowed us to unravel the enabling conditions and practices that keep rural women in agriculture from achieving sustainable livelihoods and food security for their dependents. This paradigm aided in comprehending issues of power relations between men and women in agricultural settings, and also helped illuminating the power hierarchies where study participants were involved. This approach facilitated accurate descriptions and analysis during the investigation regarding the contradictions that arise from the existence of both legal frameworks and norms governing matters concerning women's right to land in Western Kenya.

Data was collected in 2018 and the sample size was 384 obtained through an adjusted sample size formula, based on the target population of 114, 693 according to the County Government of Kakamega Website (2018). The following procedure was used:

Formula

$$S = Z \times P \times 1 - P / M^2$$

I.e. *SS*=(*Z*-score)^2 × p × (1-p)/(margin of error)^2

Z-score = 1.96 for confidence level 95%

Proportion p is your expected outcome. If you do not have any idea about the proportion, you can take 0.5. This will maximize your sample (CheckMarket, 2019).

Therefore:

$$
\begin{aligned}
S &= (1.96)^2 \times 0.5 \times (1-0.5)/(0.05)^2 \\
&= 3.8416 \times 0.5 \times 0.5 / 0.0025 \\
&= 3.8416 \times 0.5 \times 200 \\
&= 3.8416 \times 100
\end{aligned}
$$

S = 384.16 which is the sample size

To obtain an adjusted sample size of 384 for the target population of 114,693 the following formula was used:

$$
\begin{aligned}
&\text{Adjusted } S = (S)/1 + \left[(S-1)/\text{Target Population}\right](\text{CheckMarket,2009}) \\
&\text{Adjusted } S = (384.16)/)/1 + [(384.16 - 1/114{,}693] \\
&= 384.16 + [383.16/114{,}693] \\
&= 384.16 + 0.003 \\
&= 384.16(384)
\end{aligned}
$$

From a feminist perspective it was imperative that we included both women and men which aided the exploration of how change in social inequality can occur by removing obstacles and addressing power imbalances. We noted how gender differences continue to perpetuate discrimination and in turn keep women from secure access and control of land for their agricultural activities. We had to involve men in all aspects of data collection due to the fact that they are key perpetrators of harmful patriarchal practices.

To achieve a gender-sensitive sample size, we used Kakamega County gender-based percentages from the Kenya National Bureau of Statistics (KNBS), which had 48% men and 53% women. Ultimately, we multiplied the percentage of men to the sample size (384) to obtain 184 as the number of men to be included in the study. A systematic random sampling method was used to identify and select the research participants. An interval ratio of 2 was calculated by dividing the target population by the two study samples of 184 and 236 respectively. The first research participant was purposely selected and the interval of 2 was applied to identify and select subsequent research participants. Focus group discussions (FGDs) were conducted each constituting about eight persons, and we purposely included three male participants in each FGD, in meeting the gender ratios required. Twenty-one (21) interviews were done with key informants from government agencies and those from traditional governance structures,[2] and also non-governmental organizations (NGOs) in the study sites.

Results and Discussions

It is evident that Kenya has made progressive strides to achieve gender equality through legislative frameworks that seek to promote women's land rights. Emphasis is laid on the fact that agricultural development in the country cannot reach its peak without securing the rights of women to access, own, and make decisions regarding the use of land. The glaring challenge remains how to accord women their rights to capital of production in a context that is still deeply rooted in patriarchal customs that are contradictory to legal expectations.

Land ownership in the study sites was such that out of those interviewed, 97% of men interviewed owned land while 3% did not own land. Out of the women interviewed, 96% did not own land but accessed land belonging to their husbands or fathers. However, 4% of women mentioned owning land through purchase or inheritance.

Perceptions of Land Rights by Community Leaders: Legal Versus Cultural Expectations

In this study, community leaders are composed of village elders who make decisions concerning their community members and are collectively referred to as village elders. In the study sites, village elders were composed of men and women who are selected by the community. They are usually the first port of call by community members on matters of contention such as land disputes and government officials such as chiefs regularly consult them. During the focus groups, women elders were outnumbered in their views that supported women's rights to land. The conclusion from these focus groups was that land ownership was the preserve of men, and women seeking ownership was unacceptable. The basis of this was that traditions do not allow women to own land but they can access it through their husbands and fathers. It was unfathomable to the elders that a woman can own land and that their rights are protected legally. One elder was emphatic by saying:

> Wives are sojourners where they are married and that is why they cannot own land, especially ancestral land. My wife is a stranger and it doesn't therefore make sense for her to have any rights to land passed to me and my sons by our forefathers to her. (Interview during focus group discussion, East Wanga Ward, Western Kenya, 27 May 2018)

In this case, legal frameworks such as the Matrimonial Property Act, 2013, cannot protect a widow who seeks to have ownership of her late husband's land when her in-laws take over. This Act is silent on this matter and can only protect women where land was jointly acquired with their husbands.

> It is the duty of wives to live in peace with their in-laws such that when death occurs they will be allowed to access their late husband's parcel of land and property. (Interview with a male elder during focus group discussion, East Wanga Ward, Western Kenya, 27 May 2018)

When queried about their awareness of legal frameworks including the supremacy of the prescriptions of the Constitution of Kenya, 2010, regarding women's land rights, the elders were not aware. They also disputed the notion of equality of men and women as enshrined in the Constitution because it contradicts cultural norms.

Perceptions of Land Rights by Government Officials: Legal Versus Cultural Expectations

One way to ensure that women's rights to land are realized is through equal representation of men and women in land administration bodies at the grassroots. Empirical evidence collected during this study revealed that having gender balance in land administrative bodies does not equate to just decisions that would have vulnerable women access their rights to land. It was clear that both men and women members of these boards grapple with adopting the principles set out in legislative frameworks to guide their decisions. Their mindsets are still rooted in cultural norms that seek to maintain the status quo. A woman Agricultural Officer who serves in local land board had this to say:

> Yes, I believe women should be allowed to access land but this should not threaten the position of men in our society because our ancestors were not wrong in laying certain traditions in place. Why would a woman want to

> challenge her family by trying to fight for land ownership? (Interview during a focus group discussion, Malaha-Isongo-Makunga Ward, Western Kenya, 13 April 2018)

For this women Agricultural Officer, a widow or unmarried daughter seeking the board's intervention that will allow them to secure their rights to land as prescribed in law is tantamount to denying men their rightful possession. When further interrogated, she was very categorical that she would not accept land inheritance from her own father or jointly own land with her husband. On matters allowing daughters to inherit land from their fathers was contention from a more senior government official serving in the same land board—a man. He was emphatic to his decision which would keep his daughters from inheriting land from him:

> I love my daughters very much but I would not bequeath land to them because it belongs to my sons. I would expect my daughters to get married and access their husbands' parcels of land. I would not want my actions to cause any friction in their marriages because if women know that they have an alternative abode, they would not stick to their marital homes, especially when they have disputes with their husbands and in-laws. (Interview during a focus group discussion, Malaha-Isongo-Makunga Ward, Western Kenya, 13 April 2018)

With this reality in place, there is a likelihood that women seeking justice from the land board would not get the support required. There is a dire need for sensitization that helps change the mindsets of board members who will in turn be able to sensitize members of the community and deliver just decisions for vulnerable women. The clash between "Eurocentric view of patriarch" and the African view of patriarchy seems to clash in this study's setting. Pottier (2005) posits that during pre-colonial times land was communally owned and all people irrespective of gender had the right to land controlled by a political unit that they belonged to. With this mindset, it continues to be difficult to convince an African mindset that their form of "patriarchy" is oppressive and their women need to be "rescued" from it through individual tenure that Kenya's land regime espouses. Actually, the discussions alluded to a perception that the current land administration regime is likely to destroy marriages and extended family structures through according women their land rights.

When asked about the various provisions of legal frameworks that they make reference to while sorting out land cases, none of the Board members was well-articulated. In their defense they decried the lack of copies of these frameworks in their possession. Despite having smart phones, they were also not aware that access to these laws was possible through their gadgets. All is not lost because for the first time in the history of Kenya women are now being incorporated in land boards, a situation that did not happen before the new regime of laws when the boards were referred to as Land Control Boards. It was evident that NGOs and government had their work cut out in terms of changing mindsets.

It is also important to note that population increase has also placed pressure on finite land in the region, a situation that has led men to want to hold on to the little parcels of land that remains. When their wives, daughters, and sisters seek a share of this limited land, the men are bound to "fight" for what is "rightfully theirs" through cultural norms.

Perceptions of Land Rights by Women: Legal Versus Expectations

It was a common response from women in group discussions that women were expected to be married off and go ahead to "own" land with her husband. Below is an excerpt from women's discussion:

> Women are creators of relations and friendships so they are expected to leave their fathers' land and go out and be joined to another family [...] she cannot therefore, make claim to her fathers' land mainly because she is not allowed to bring her husband to live on her father's land. (Focus Group, 2018)

With this type of response, it was evident that women's rights to land, especially ownership, were not guaranteed, and this was expected. When interrogated about the rights of unmarried women or widows, it was clear that such women are "allowed" to access land for agriculture upon the agreement of in-laws, fathers, or extended male members of their families. Daughters and divorced women can only access land through consent from their fathers. This consent is not guaranteed and sometimes the quality of land allocated to such women is necessarily the best. A widow who had gone back to her ancestral home did not have a good experience because the land she was allocated was not of good quality (Focus Group, 2018)

> I lost access to my late husband's land when my in-laws treated me in an unwanted manner, mainly because I did not bare my husband, sons. When I went back to my father's land my brothers and uncles apportioned me land that is not the most fertile and far from the stream. But, I am better off compared to some women I know who have not been allowed to settle back on ancestral land.

Such responses demonstrated that the women were not aware of their rights that are enshrined in the Constitution, which speak against any form of discrimination based on gender in as far as property rights are concerned.

There were cases where women interviewed revealed that their efforts to own land through purchase were hampered by their husbands who insisted that titles be on their (husband's) name. When questioned on their knowledge on legal frameworks that protect their rights, specifically joint-ownership, for example the Matrimonial Property Act, 2013, the general perception was that culture supersedes laws. On the contrary, the majority were not aware that the Constitution of Kenya, 2010, and other legislations override cultural norms.

During an FGD in East Wanga Ward, there was a case of a widow who was able to articulate knowledge on her rights to land. Her knowledge did not suffice in helping her gain control and ownership of land that belonged to her husband:

> I lost my husband who left me with sons and daughters. My brother-in-law continues to harass me because he seeks to inherit my husband's parcel of land. He has even gone ahead to move the boundaries that had been long allocated by my father in-law before he passed on. I know I can pursue justice through legal means but it costs a lot of money to hire an advocate. I also live too far from town where I need to access help. I cannot afford those trips. I sought help in locating important documents from the Ministry of Lands, a move that proved futile because the files went missing the second time I went back to do follow up. My only hope is when my sons grow up they can fight for their rights. (Interview during focus group, East Wanga Ward, Western Kenya, 16 April 2018)

There was a glimpse of hope when a woman respondent in Malaha-Isongo-Mukunga Ward narrated that her husband took the initiative of doing joint-titling as he is not on good terms with his brothers:

> My husband sensed negative scenarios play(ing) out in his family where land was taken away from widows and orphans, and decided to follow up on joint-titling when succession was done amongst him and his brothers'. (Interview, Malaha-Isongo-Makunga Ward, Western Kenya, 16 April 2018)

With such a scenario cropping up, it is clear that men are not static in their thoughts and that they can be influenced to make decisions that protect the women in their lives. To achieve a change of mind among men would require a removal of fear that they have of women who are empowered through land rights. During an FGD in Namamali Ward, one male responder laid his thoughts as follows:

> A marriage cannot have two men. When women have the same rights as men in land ownership and control, they are bound to be unruly. They become tough headed and stop listening to their husbands. We are aware that NGOs are educating our women on what the law says about their rights and the "little" knowledge they get makes them feel equal to us. We will never be equals. (FGD, Namamali Ward, Western Kenya, 12 June 2018)

Conclusion

Ownership of land by women will guarantee that women have access to capital of production for agriculture. However, it is seemingly clear that women's ownership of land will take a long time. Overall there seemed to be a tug of war between law and culture because the community members (both men and women) still hold on tightly to their cultural norms. Nonetheless, hope is not lost because this study uncovered instances of co-ownership and individual ownership of land by women. These cases need to be highlighted for the benefit of all, and in the meantime the country needs to focus on strategic awareness creation and research that tells the benefits of affording women their constitutional rights. Informal learning is one of the ways that various stakeholders including traditional community leaders can be sensitized on the legal framework. This kind of learning should endeavor to understand cultural norms and debunk fears that men have about empowered women. Judicial processes in the country need to find ways to remove the challenges that women who are aggrieved by customary practices regarding their land rights have to go through to get justice. Information required by such women should be cascaded to local administrative structures such as land boards and chief offices for easier access. Land board members' capacities need to be enhanced in as far as knowledge of legal frameworks and what is expected of them. It is also important that expectations are set for these land boards and outcomes of their decisions evaluated to ensure they are justly executing their roles.

Women are slowly learning about their rights and these few enlightened cases need to be used to educate the community. These women made the decision to acquire land mainly without their knowledge to carry out their agricultural activities because of limitations they experienced in their matrimonial homes. Their decisions were informed by the knowledge that they have such rights under the law, and the desire to secure themselves in case they lost their matrimonial rights over land. Additionally, there are reasons behind the majority of men's decision to stick to culture and this study perceived it to be an issue of power relations and gender. It seemed that holding on tightly to culture made the men feel in control within their households. This is after it was severally revealed by male respondents that women who have property are "unruly" and they could easily leave their matrimonial homes because they have an alternative on account of owning land.

Notes

1. Protocol to the African Charter on Human and Peoples' Rights on the Rights of Women in Africa, adopted by the 2nd Ordinary Session of the Assembly of the Union, Maputo, September 13, 2000, CAB/LEG/66.6, entered into force on November 25, 2005, art. 7.
2. Ministry of Land and Physical Planning, Ministry of Agriculture and Irrigation, judiciary, sub-county land boards, assistant chiefs, and village elders.

References

Anukriti, S. (2014). The Fertility-Sex Ratio Trade-Off: Unintended Consequences of Financial Incentives (No. 8044). IZA Discussion Paper.

Ardovini-brooker, J. (2002). Feminist Epistemology: The Foundation of Feminist Research and its Distinction from Traditional Research. *Advancing Women in Leadership*, 10.

Bikketi, E., Speranza, C. I., Bieri, S., Haller, T., & Wiesmann, U. (2016). Gendered Division of Labour and Feminisation of Responsibilities in Kenya; Implications for Development Interventions. *Gender, Place & Culture, 23*, 1432–1449. https://doi.org/10.1080/0966369X.2016.1204996

Burke WJ. Jayne TS. 2014. Smallholder Land Ownership in Kenya: Distribution between Households and Through Time. Agricultural Economics, 45: 185–198. dooi:https://doi.org/10.1111/agec.12040.

CheckMarket. (2019). *Optimal Sample Size*. CheckMarket. https://www.checkmarket.com/kb/calculate-optimal-sample-size-survey/. Accessed April 12, 2019.

County Government of Kakamega. (2018). *Sub Counties – County Government of Kakamega*. Available at: https://kakamega.go.ke/category/aboutus/political-units/sub-counties/. Accessed March 7, 2018.

Davis, B., Di Guiseppe, S., & Zezza, A. (2017). Are African Households (not) Leaving Agriculture? Patterns of Households' Income Sources in Rural Sub-Saharan Africa. *Food Policy, 67*, 153–174.

Diiro, G. M., Seymour, G., Kassie, M., et al. (2018). Women's Empowerment in Agriculture and Agricultural Productivity: Evidence from Rural Maize Farmer Households in Western Kenya. *PLoS One, 13*, 1–27.

FAO. (2020). *Protecting and Promoting Women's Land Rights in the Face of COVID-19 and Beyond*. Available from https://www.fao.org/partnerships/parliamentary-alliances/news/news-article/en/c/1310413/. Accessed July 15, 2022.

Galiè, A., Mulema, A., Benard, M. A. M., Onzere, S. N., & Colverson, K. E. (2015). Exploring Gender Perceptions of Resource Ownership and Their Implications for Food Security Among Rural Livestock Owners in Tanzania, Ethiopia, and Nicaragua. *Agriculture & Food Security, 4*, 2. https://doi.org/10.1186/s40066-015-0021-9

Government of Kenya. (2010). *The Constitution of Kenya, 2010*. National Council for Law Reporting. Available from http://cn.invest.go.ke/wp-content/uploads/2018/10/The-Constitution-of-Kenya-2010.pdf. Accessed July 13, 2022.

Government of Kenya. (2012). *The Land Act, 2012*. Kenya Law Reform.

Government of Kenya. (2013). *Matrimonial Property Act*. National Council for Law Reporting.

ILO. (2019). *Employment in Agriculture, female (% of female employment)-(Modeled ILO estimate- Kenya*. Available from https://data.worldbank.org/indicator/SL.AGR.EMPL.FE.ZS?locations=KE. Accessed July 7, 2022.

Kameri-Mbote, P., & Muriungi, M. (2018). *Assessing Women's Land Rights and Food Security in Kenya: Challenges and Opportunities*. Annual World Bank Conference Land and Poverty.

Kansiime, M. K., Girling, R. D., Mugambi, I., Mulema, J., Oduor, G., Chaha, D., Ouvrard, D., Kinuthia, W., & Garratt, M. P. M. (2021). Rural Livelihood Diversity and Its Influence on the Ecological Intensification Potential of Smallholder Farms in Kenya. *Food and Energy Security, 10*, e254. https://doi.org/10.1002/fes3.254

Karanja, P. W. (1991). Women's land ownership rights in Kenya. *Third World Legal Studies, 10*(6), 109–135.
Kihiu, E., & Munene, B. (2021). *Women's Access to Agricultural Finance in Kenya*. KIPPRA Policy Brief No. 03/2020-2021.
Landesa. (2014). *Women's Land and Property Rights in Kenya*. Available from: https://www.landesa.org/wp-content/uploads/LandWise-Guide-Womens-land-and-property-rights-in-Kenya.pdf. Accessed July 13, 2022.
Mbugua, S. (2018). *Despite New Laws, Women in Kenya Still Fight For Land Rights*. The New Humanitarian. Available from https://deeply.thenewhumanitarian.org/womensadvancement/articles/2018/02/23/despite-new-laws-women-in-kenya-still-fight-for-land-rights. Accessed June 18, 2022.
Meillassoux, C. (1981). *Maidens, Meals and Money: Capitalism and the Domestic Community. Themes for Social Sciences* (pp. 61–78). Cambridge University Press.
Mishra, K., & Sam, A. G. (2016). Does Women's Land Ownership Promote Their Empowerment? Empirical Evidence from Nepal. *World Development, 78*, 360–371.
MoALF. (2017). *Climate Risk Profile for Kakamega County*. Kenya County Climate Risk Profile Series.
Munala, P. N. (1995). *Property Ownership among the Luhya women of Western Kenya*. MA Thesis, University of Nairobi, Kenya.
Musangi, P. (2017). *Women Land and Property Rights in Kenya. Paper Prepared for Presentation at the World Bank Conference on Land and Poverty, 20–24 March, 2017*. The World Bank.
Onyalo, P. O. (2019). Women and Agriculture in Rural Kenya: Role in Agricultural Production. *International Journal of Humanities, Art and Social Studies, 4*, 4.
Oya, C. (2011). Contract Farming in Sub-Saharan Africa: A Survey of Approaches, Debates and Issues. *Journal of Agrarian Change, 12*, 1–33.
Palacios-Lopez, A., Christianen, L., & Kilic, T. (2017). How much of the labor in African agriculture is provided by women? *Food Policy, 67*, 52–63.
Pottier, J. (2005). 'Customary Land Tenure' in Sub-Saharan Africa today: Meanings and contexts. In C. Huggins & J. Clover (Eds.), *From the Ground Up: Land Rights, Conflict and Peace in Sub-Saharan Africa* (pp. 55–75). Pretoria.
The County Government of Kakamega. (2013). *Kakamega County Integrated Development Plan: 2013–2017*. Government Printers. Available at: http://www.busiacounty.go.ke/?p=2098.
The International Women's Human Rights Clinic (IWHRC). (2008). *Women's Land and Property Rights in Kenya-Moving Forward into a New Era of Equality: A Human Rights Report and Proposed Legislation*. Georgetown University Law Center.
World Bank. (2011). *In Kenya, Survey of Female Farmers Uncovers Challenges*. World Bank Feature Story. Available from https://www.worldbank.org/en/

news/feature/2011/09/09/in-kenya-survey-of-female-farmers-uncovers-challenges. Accessed July 8, 2022.

Yamano, T., Place, F. M., Nyangena, W., Wanjiku, J., & Otsuka, K. (2009). Efficiency and Equity Impacts of Land Markets in Kenya. In S. T. Holden, K. Otsuka, & F. M. Place (Eds.), *The Emergence of Land Markets in Africa: Impacts on Poverty Equity and Efficiency*. Resource for the Future. isbn 978-1-933115-69-6.

CHAPTER 2

Representation of Women and Gendered Role Portrayals in Television Advertising: A Content Analysis of Ramadan Advertising in the Pre- and Post-Tunisian Revolution

Mouna Ben Ghanem and Fawzi Dekhil

Introduction

Over the last few decades in Tunisia, we have noticed important developments in marketing communication, especially media-based communication. Spending on media communication has risen from 8 million Tunisian Dinars in 1982 to 100 million Tunisian Dinars in 2007 to 219 million Tunisian Dinars in 2017 (Mediascan[1]). In Tunisia, TV advertising was launched in 1987. From 17 million Tunisian Dinars in 2000, investments in televised advertising reached 35 million Tunisian Dinars in 2007 to 135 million Tunisian Dinars in 2017.

M. B. Ghanem (✉)
University of Tunis El Manar, Tunis, Tunisia

F. Dekhil
Faculty of Economic Sciences and Management, University of Tunis El Manar, Tunis, Tunisia

B. S. Nayak (ed.), *Political Economy of Gender and Development in Africa*, https://doi.org/10.1007/978-3-031-18829-9_2

A TV advertisement is a comprehensive message that comprises not only words, but also images, sounds, and movements (Belch et al., 2004; Décaudin, 2003). This form of advertising has been the subject of several studies aimed at evaluating its efficiency, structure, and creation strategies. Overall, two main research approaches emerge from the literature: the first is approach related to the analysis of the efficiency of TV advertising among consumers (Shrum et al., 1998; Taylor et al., 1997) and the second is related to the analysis of the advertisements' content (Gilly, 1988; Lerman & Callow, 2004; Millner & Higgs, 2004; Resnik & Stern, 1977). A number of studies focused on the content analysis of gender role portrayals (Ferguson et al., 1990; Jeryl & Jackson, 1997; Millner & Higgs, 2004; Wee et al., 1995a, 1995b, 1995c; Zhou & Chen, 1997).

The objective of this study is to explore the portrayals and roles of men and women in TV advertising in the Tunisian context. Several studies have shown the importance and efficiency of using human characters in advertising (Dupond, 1993; Gavard-Perret, 1993). Many scholars have posed the following question: Do the portrayals and roles of men and women in advertising reflect reality? If so, why? And if not, why not? (Gilly, 1988; Pollay, 1986). Schneider and Schneider (1979) state that advertising should broadly reflect the cultural norms and certain stereotypes of men and women. This observation motivates us to conduct this study in Tunisian content. Several similarities and differences in gender role portrayals in ads have been noted across cultures (Das, 2000; Wiles et al., 1995). For this, it is important to study gender role portrayals in a larger number of countries. This study explores the evolution of representations of male and female roles in Tunisian television advertising in the light of the potential influence of the political, economic, social, and cultural changes that have occurred in Tunisia.

Another motivation for our study is that Tunisia has undergone radical changes especially with the events of the Arab Spring, which have in fact been ignited by Tunisia and the Tunisian revolution. Therefore, the current study examines gender role portrayals in advertising before the Tunisian revolution specifically during the month of Ramadan in 2007 and after the Tunisian revolution specifically during the month of Ramadan in 2017, in order to determine whether they reflect or not the cultural changes of Tunisian society in these years. In other words, it is important to examine whether cultural portrayals in Tunisian advertising have kept pace with societal changes. By reflecting on certain gender roles, Tunisian advertising reinforces the belief that those roles are best and proper. For example, if women are improperly presented in Tunisian advertisements, it is detrimental to Tunisian society because it perpetuates misconceptions. For this, we indicate that men and women portrayals construct a specific portrait of reality, and viewers adopt attitudes and expectations about the world that coincided with their vision. According to the Cultivation Theory, target receivers who are exposed to a particular view of the world in the media are beginning to accept this world as a reality (Fullerton & Kendrick, 2000), which implies a responsibility for advertisers. Thus, it is important to examine and understand similarities and differences in gender role portrayals in the years 2007 and 2017.

To the best of the authors' knowledge, this is the first study that attempts to explore gender role portrayals before and after the Tunisian revolution. Few countries in the Arab have experienced such historical and socio-political transformation in these years. From a political point of view, Tunisia has moved from a dictatorship to a full democracy. During this period, a substantial cultural change starts to take place in the predominant values attached to industrialized societies: rationalism, pragmatism, tolerance, egalitarianism, willingness to participate in decision-making processes, and a penchant for consumer spending as mentioned by Lo'pez Pintor (1990).

Context of the Study

Creation in advertising requires the creators to use a variety of advertising methods and styles. In their ads, they incorporate various characters (men and/or women, babies and/or teenagers and/or adults, experts, celebrities, ordinary people, etc.). These characters communicate verbally through words and non-verbally through facial expressions, gaze, posture, gestures, and a smile. Incorporating human characters makes the ad more understandable, easier to interpret, and provides a sense of humanity. Advertising is also about staging characters, who are made to play roles and display attitudes and behaviors (Gavard-Perret, 1993). These people enable the desired advertising message to be transmitted and make the advertising more realistic and interactive with the recipient. Human presence plays an important role in the effectiveness of the advertising message since inserting characters into a TV advertisement makes the announcement's presentation more concrete and realistic. It attracts the recipient's attention toward the ad and increases its memorization rate, as compared to an ad devoid of human presence.

The feminine image has undergone major changes in society as well as in advertising. It has evolved from a traditional image which represented women as inferior to men, to a modern image which reflects a degree of self-awareness. Feminists and researchers have denounced the role assigned to women in advertising, which creates and reinforces negative perceptions as well as stereotypes concerning the place of women in society (Gilly, 1988). Indeed, the traditional image of women's role has been criticized by many researchers. Many scholars (A. E. Courtney & Lockeretz, 1971; Dominick & Rauch, 1972; Ford et al., 1994; Manstead & Culloch, 1981; MCArthur & Resko, 1975; O'Donnell & O'Donnell, 1978) have shown that the woman has always been presented as a housewife, a mother, or a sexual or decorative object. But for the last twenty years the image of women has been present not only in the traditional realm (home, mother, grandmother, etc.) but also in the field of active life. As a matter of fact, the researchers (Ferguson et al., 1990; Mays & Brady, 1991) have stated that the portrayal of women in advertising has changed: women are now less often featured as housewives and more as individuals who can provide better information about the product. Klassen et al. (1993) have concluded that the number of advertisements showing traditional women has dropped, while the number of those representing equality between men and women has risen.

Besides, Ford et al. (1991) have demonstrated that the perception of society has an important role in criticizing the representation of women in advertising. This concern is based on the perception that society is influenced by the representation of women's roles in advertising. In other words, a faulty description of women in advertising is harmful to society, because it creates or perpetuates false ideas. In this regard, Ford et al. (1999) have examined the development of female conscience and activity. The portrayal of women and men alters according to their representation and their roles as they appear in TV advertising. Over the years, most studies on the role of the portrayals of women and men have led to significant results (see, e.g., Gilly, 1988; Wee et al., 1995a, 1995b, 1995c; Jeryl & Jackson, 1997; Siu & Au, 1997; Millner & Higgs, 2004).

The Relationship between the Presence of Women and Men in TV Advertising and the Type of Product Promoted in the Advertisement Several studies have been conducted on this topic. A study by Millner and Higgs (2004) shows that there is a statistically significant relationship between gender and the products targeting gender. They found that in 23.8% of the advertisements, women appeared presenting feminine products and that in 75.4% they appeared with products targeting both men and women. In 4.2% of the ads, however, men promoted male products, and in 92.4% of the ads, they appeared for both men and women. This result is similar to that emerging from the Gilly (1988) study of American and Mexican advertising, and those by Wee et al. (1995a, 1995b, 1995c) and Siu and Au (1997). However, Gilly (1988) found that there is no statistically significant link between gender and advertisements targeting gender in Australian advertisements. Indeed, 7.7% of the ads presented women for products targeting women, while no woman appeared for male products; in 92.3% of the ads, women presented products targeting both men and women. As for men, in 4.1% of the ads they appeared for products aimed at women and for others aimed at men, and in 95.9% of the ads they appeared for products aimed at both men and women. Hence, we hypothesize the following: ***H1:*** *Men and women appear for products targeting both men and women.*

The Voice-Over Gilly (1988) found that masculine voice-overs are predominant (67.9%). This finding is similar to those of Wee et al. (1995a, 1995b, 1995c), Jeryl and Jackson (1997), Siu and Au (1997) and Millner and Higgs (2004), who have shown that masculine voice-overs predominate. We, therefore, hypothesize the following: ***H2:*** *Masculine voices are more commonly used than feminine ones in TV advertising.*

The Relationship between the Presence of Women and Men in TV Advertising and the Shooting Location Siu and Au (1997) show that there is a statistically significant relationship between gender and shooting location in advertising in Singapore. In fact, 54.6% of advertisements featured women at home, while 33.9% of them featured men outdoors. This result coincides with the finding of the study of Wee et al. (1995a, 1995b, 1995c) in Malayan advertising, on the RTM1 (TV3) channel, and with the result of the Millner and Higgs (2004) study. However, the study of Wee et al. (1995a, 1995b, 1995c) in Singapore advertising on the SB5 channel, and that of Siu and Au (1997) in Chinese advertising, showed that there was no significant relationship between gender and shooting location. Besides, Jeryl and Jackson (1997) identified a statistically significant link between gender and shooting location in French and British advertising: women are featured at home, while men are shown at work. This result is similar to that of the Gilly (1988) study on American advertising. Yet, Gilly (1988) shows that there is no significant relationship in Mexican and Australian advertising, where both men and women appear at home. Thus, the following hypothesis is advanced: ***H3****: Women appear at home, whereas men appear outdoors.*

The Relationship between the Presence of Women and Men in TV Advertising and Age Researchers have shown that there is a statistically significant relationship between gender and age. For example, Millner and Higgs (2004) found that 59% of the Australian ads presented young women, while 53.1% presented young men. This finding is in keeping with the observations of Gilly (1988) and Jeryl and Jackson (1997). Accordingly, we propose the following hypothesis: ***H4****: Women appearing in ads are younger (under 35) than men.*

The Relationship between the Presence of Women and Men in TV Advertising and Family Status Prior studies have shown that there is no statistically meaningful relationship between family status and the characters featured in advertisements. For example, Millner and Higgs (2004) found that 56.6% of ads featured women, 54.6% featured men, but in none of these ads there is an indication of their family status. In 32% of the ads, women are presented as married, and in 33.6% of them men are presented as married. This finding is similar to those of Gilly (1988), Wee et al. (1995a, 1995b, 1995c), Jeryl and Jackson (1997), and Siu and Au (1997). We, therefore, hypothesize the following: ***H5****: While a few characters have appeared as married, there is no indication of the family status of the majority of men and women.*

Concerning the Job of the Character Featured in an Advertisement Millner and Higgs (2004) show that there is a statistically significant relationship between the job and the character appearing in an ad. In fact, they found that 67.2% of the ads presented women, and 48.7% presented men with no indication of their jobs. Women are not in a working situation in 21.1% of the ads, while men are in a working situation in 32.8%. This result is similar to that in the Gilly (1988) study of American and Mexican advertising, and to those of Wee et al. (1995a, 1995b, 1995c) and Jeryl and Jackson (1997). However, Gilly (1988) shows that in Australia, there is no statistically significant relationship between jobs and the characters featured in advertisements. In fact, 63.8% of the ads presented women and 52.1% presented men, with no indication as to their work; 36.2% of the ads presented women in non-working situations. Men, however, are presented in a working situation in 41.7% of cases. This finding is similar to what was found by Siu and Au (1997). Accordingly, we propose the following hypothesis: ***H6**: There is no indication of the jobs of the majority of men and women. Women generally appeared as not working, while men appeared in working situations.*

Concerning the Positions Occupied by the Characters Featured in Advertising Jeryl and Jackson (1997) show that there is a statistically significant link between the characters in ads and their positions occupied in French and British advertising, in which men appear as executive managers, and women as housewives. This is in line with the studies of Gilly (1988) on Mexican advertising, and Millner and Higgs (2004). However, the study by Wee et al. (1995a, 1995b, 1995c) testifies to the absence of any significant difference between advertising characters and positions occupied, which is in keeping with what Gilly (1988) found in the USA and Australia. Moreover, the work of Siu and Au (1997) demonstrates a significant relationship between positions occupied and the advertising characters in Singapore and China. In fact, women are featured as office workers and men as executive managers. We, therefore, hypothesize the following: ***H7**: Women appeared as housewives, men as executive managers.*

Concerning the Relationship between the Spokesperson and the Characters Featured in Advertising In their studies, Siu and Au (1997) show that in Singapore advertising there is a statistically significant relationship for the characters who appeared as spokespersons for a product or a service: 34.2% of the ads show women as spokespersons for the product or service, while 23.2% have men as spokespersons. This result is in line with the study by Jeryl and Jackson (1997) on British advertising. However, in their studies, Gilly (1988), Wee et al. (1995a, 1995b, 1995c), Siu and Au (1997), Jeryl and Jackson (1997), and Millner and Higgs (2004) show that there is no statistically significant relationship when characters appeared as spokespersons for the product or the service. Therefore, the following hypothesis is advanced: ***H8**: Both men and women appeared as spokespersons for the product or service in advertisements.*

Concerning the Relationship between the Spokesperson's Credibility and the Characters Who Appeared in TV Ads Jeryl and Jackson (1997) show the existence of a significant relationship regarding the credibility of the spokesperson of the characters featured in French and British advertising. In France, in 94% of ads women are spokespersons when they were users of the product. Men, however, were spokespersons when they were not users of the product in 43% of ads. In British advertising, 79% of the ads featured women as spokespersons when they were users of the product, whereas men were spokespersons when they were not users of the product in 48% of the ads. This finding is similar to that seen in the Gilly (1988) study of American and Mexican advertising, and in the Siu and Au (1997) study of Chinese advertising. However, the Gilly (1988) study shows no significant credibility difference in the gender spokesperson in Australian advertising, which is in keeping with the Siu and Au (1997) study of Singapore advertising, and with Wee et al. (1995a, 1995b, 1995c) and Millner and Higgs (2004). We, therefore, propose the following hypothesis: ***H9****: Women were spokespersons when they were users of the product, while men were spokespersons when they were not users of the product.*

In Terms of "Help" for the Character Featured in TV Advertising Gilly (1988) shows that there is a statistically significant link between help and the characters featured in Mexican advertising: 88.3% of ads featured women who had no help; only in 8.3% of ads they appeared as recipients of help, which also is the case for men in 87.3% of cases; only in 9.8% of ads they appeared as help providers. Millner and Higgs (2004) show in their studies that there is a significant difference between gender and aid in Australian advertising. Indeed, 88.5% of ads showed women who were not helped; only in 9.0% of ads they appeared as help providers. While 93.3% of ads showed men without any help, only in 3.4% of ads they appeared as help providers and also as help recipients. However, the Gilly (1988) study testifies to the absence of any significant difference between gender and help in American and Australian advertising. This result is similar to those obtained by Wee et al. (1995a, 1995b, 1995c), and Jeryl and Jackson (1997). We, therefore, propose the hypothesis below: ***H10****: There is no difference in the amount of help men and women receive in the advertisements.*

Concerning "Advising" the Character Featured in Advertising Wee et al. (1995a, 1995b, 1995c) find a statistically significant relationship between advice and the characters appearing in advertisements in Singapore and Malaysia, on the RTM1 channel: 82.8% of the ads in Singapore featured women who were not given any advice; only in 9.5% of ads they appeared as counseling recipients. While 96.2% of the ads in Singapore featured women who were not given any advice, only in 1.9% of ads they appeared as providers of counseling and also as counseling recipients. On the TRM1 channel in Malaysia, 84.7% of ads showed women who were not given any counseling; only in 10.5% of ads they appeared as counseling recipients. While 98.2% of ads showed men who were not given any counseling, only in 1.9% of ads they appeared as providers of counseling. This result is similar to Gilly (1988) findings on American advertising. In addition, Millner and Higgs (2004) show a significant difference between gender and counseling in Australian advertising. Indeed, 91% of the spots showed women who were not given any counseling; only in 9% of ads they appeared as providers of counseling. While 96.6% of the spots showed men who were not given any counseling, only in 2.5% of ads they appeared as providers of counseling. However, Gilly (1988) found no significant relationship between gender and counseling in Mexican and Australian advertising. This result is similar to that obtained in the study by Wee et al. (1995a, 1995b, 1995c), on the TV3 channel, and in the studies by Jeryl and Jackson (1997) and Millner and Higgs (2004). We therefore hypothesize the following: ***H11**: There is no advice between men and women in advertisements.*

Concerning the Relationship between the Presence of Women and Men and Their Role as it Appeared in Advertising Several studies examined this relationship. For instance, Millner and Higgs (2004) show a significant role for characters featured in TV advertising. They found that 52.5% of ads presented women in roles which kept them independent from others, which was the case for men in 63% of ads. However, Jeryl and Jackson (1997) found that in British advertising 32% of advertisements presented women in roles in which they were independent from other people, which was the case for only 25.5% of men. In the United Kingdom, women were dependent in 33% of ads, while men were dependent in 18% of ads. This finding is similar to what was reached by Gilly (1988), Wee et al. (1995a, 1995b, 1995c), and Siu and Au (1997). We, therefore, state the following hypothesis: ***H12****: Women as well as men have appeared in roles in which they are independent of others.*

In Terms of Activity Millner and Higgs (2004) showed a statistically significant relationship for characters who were performing a physical activity in Australian advertising. They found that women appeared as inactive in 96.7% of ads, while men appeared inactive in 84% of ads. This result is similar to that observed by Siu and Au (1997). However, Gilly (1988), Wee et al. (1995a, 1995b, 1995c), and Jeryl and Jackson (1997) showed that there was no statistically significant relationship for the characters who had a physical activity in advertisements. Hence, we propose the following hypothesis: ***H13****: Both women and men have appeared as inactive.*

In Terms of Frustration Gilly (1988) found a statistically significant relationship between frustration and the characters featured in Mexican advertising. According to Gilly (1988), 98.9% of ads presented women as having *an unrelieved frustration*, which was the case for men in 98% of ads. However, Gilly (1988) demonstrated the existence of a statistically significant relationship between frustration and the characters featured in American and Australian advertising. This agrees with the findings of Wee et al. (1995a, 1995b, 1995c), Jeryl and Jackson (1997), and Millner and Higgs (2004). Hence, the hypothesis is advanced: ***H14****: Women and men appeared as having an unrelieved frustration.*

In Terms of Social Role In terms of social role of the character appeared in advertising, Sexton and Haberman took into consideration the social role of gender appeared in society. The traditional role: the situation in which gender appears in advertising would be considered by society as traditional, and the non-traditional role: the situation in which gender appears in advertising would be considered by society as non-traditional. Hence, the following hypothesis is advanced: ***H15****: Women and men appeared in television commercials in a modern role.*

In Terms of Religious Allusion *There is a statistically significant relationship between gender and religious allusion ($p = 0.000 < 0.05$). Therefore, hypothesis H16 is retained.*

Method to Study Gender Roles and Portrayals

The creative professional working in an advertising agency engraved the commercials televised appeared during the month of Ramadan in 2007 and 2017 on a CD-ROM. The difference in the timing of data collection in Tunisia is taken into account before and after the Tunisian revolution. We carried out a content analysis on 201 Tunisian ads, among a set of 613 ads in 2007. We also carried out a content analysis on 291 Tunisian ads, among a set of 304 ads in 2017. We first discarded both the duplicates (for different "formats": 45, 30, 20, 10, and 5 seconds), as in the studies of Manstead and Culloch (1981), Walliser and Moreau (2000), and Wee et al. (1995a, 1995b, 1995c), and also the ads that did not use human characters (Schneider & Schneider, 1979). The ads were broadcasted during the month of Ramadan in 2007 and 2017, between 5 and 9 p.m. (prime time) on three channels: *Canal 7, Canal 21*, and *Hannibal TV*. Two elements accounted for the choice of the month of Ramadan: the large audience for the three channels and the TV advertising spending, which rises during this month.

Content Analysis In this research, TV advertisements were analyzed by the content analysis technique. This method allows the user "to identify the similarities and the differences within an advertising content, in terms of the roles through which women and men are represented" (Lerman & Callow, 2004). It "also allows to classify the text or the objects in predetermined categories the objective of which is to compare their basic components (that is the content). This technique enables us to determine the quantity of the qualitative data through the monitoring of the presence or the frequency of a word or an object".

Analysis Grid and Measuring Variables The measuring instrument for our research is an analysis grid which was built by gathering several analytical variables used by a number of researchers, including Gilly (1988), Jeryl and Jackson (1997), Siu and Au (1997), and Millner and Higgs (2004). In this grid, two groups of analytical variables are considered: **Group 1: Global variables:** They are used to analyze each advertisement in terms of product, product user, sound comments, and shooting location.

Group 2: Gender variables: They are used to analyze each advertisement in terms of the demographic variables of the featured character (gender, age, family status, employment, and occupation); and approach variables (spokesperson, spokesperson's credibility, help, advice, role, activity, frustration). Some of these variables are defined in Appendix 1.

The Coding Process The variables used in the analysis were coded for each character appearing onscreen; this is consistent with the studies of Gilly (1988), Jeryl and Jackson (1997), Millner and Higgs (2004), Schneider and Schneider (1979), Siu and Au (1997), and Wee et al. (1995a, 1995b, 1995c). The coding of variables was performed by two researchers. When an independent evaluation of each ad was completed, the coding data were compared, and disagreements were resolved by discussion. Thereafter, the most logical reasoning was accepted.

Statistical Analysis Data were analyzed using IBM SPSS 23. Descriptive statistics are presented in Appendix 2. We then performed a crossed sorting between the featured character's gender variable and the descriptive, demographic, and approach variables. The findings are based on the Chi-square test and the analysis of a one-factor variance (ANOVA) (see Appendices 3 and 4).

Preliminary Analysis of TV Ads in 2007 and 2017

In 2007 (2017), 51.7% (38.1%) of the TV ads were shot outdoors, 32.3% (23.4%) of them indoors, 10.4% (19.6%) at work, 5.5% (3.8%) in a shop, and 0.0% (15.1%) no shooting location. Among the categories of products presented in these ads, 45.3% (48.8%) are services, 39.8% (28.2%) are food products, 9.5% (2.1%) are cosmetics, 3.0% (6.9%) are cleaning products, and 2.5% (14.1%) concern entertainment products. Among them 89.6% (94.5%) of the products promoted in the ads target both men and women, while 5.0% (1.7%) target women, 3.0% (2.7%) are aimed at men, and 2.5% (1.0%) at children. In 56.7% (30.2%) of ads, spoken comments are made by a man's voice. A woman's voice is used in 20.9% (40.2%) of the ads. While 89.6% (74.9%) of the ads studied feature characters adopting a modern social role, only 10.4% (4.5%) of them show traditional roles.

Results

Analysis of the Portrayals of Men and Women

To perform a content analysis of gender role portrayals, we used a crossed sorting of gender and descriptive variables, demographic variables, and approach variables (see Appendices 2 and 3).

Descriptive Variables
At the level of descriptive variables, we found a statistically significant relationship between gender and ***products targeting gender*** ($p = 0.000 < 0.05$). Therefore, hypothesis 1 is supported. In fact, in 2007 (2017), 77.8% (88.5%) of ads showed that women appeared alone to present products aimed at both men and women. In 16.7% (9.6%) of ads, women appeared to present products designed for women, 5.6% (0.0%) of ads to present products designed for children, and 0.0% (1.9%) of ads to present products designed for men. As for men, they appeared alone presenting products intended for both men and women in 87.5% (92.6%) of ads and intended for men only in 12.5% (7.4%) of ads. When they appeared together, men and women presented products intended for both men and women in 93.2% (97.6%) of ads, products targeting women in 3.0% of ads, products for children in 2.3% (2.4%) o of ads, and masculine products in 1.5% (0.0%) of ads. This is in keeping with the studies conducted by Gilly (1988), Wee et al. (1995a, 1995b, 1995c), Jeryl and Jackson (1997), Siu and Au (1997), and Millner and Higgs (2004). We also found a statistically significant relationship between gender and ***product category*** ($p = 0.001 < 0.05$). In 2007 (2017), women appeared alone in 61.1% (40.4%) of ads to present food products and in 22.2% (57.7%) to present services. Men appeared alone in 59.4% (44.4%) of ads to present services and in 18.8% (31.5%) to present food products. When appearing together, men and women presented services in 48.1% (52.8%) of ads and food products in 39.1% (24.0%).

Concerning the soundtrack, there is a statistically significant relationship between gender and ***voice-over*** ($p = 0.000 < 0.05$). Masculine voice-overs prevail in 81.3% of ads. This is in keeping with the studies conducted by Gilly (1988), Wee et al. (1995a, 1995b, 1995c), Jeryl and Jackson (1997), Siu and Au (1997), and Millner and Higgs (2004). While, in 2017, feminine voice-overs prevail in 59.6% of ads. This is contrary in keeping with the studies conducted by Gilly (1988), Wee et al. (1995a,

1995b, 1995c), Jeryl and Jackson (1997), Siu and Au (1997), and Millner and Higgs (2004). Therefore, hypothesis 2 is partially supported.

In addition, there is a statistically significant relationship between gender and the *shooting location of the advertisement* ($p = 0.005 < 0.05$). Therefore, hypothesis 3 is supported. In 2007, women appeared alone, at home, in 50% of ads. Men appeared alone but outdoors in 65.6% of ads. Men and women appeared together, outdoors, in 51.1% of ads. This is consistent with the studies conducted by Wee et al. (1995a, 1995b, 1995c), Siu and Au (1997), and Millner and Higgs (2004). In 2017, women appeared alone, at home, in 40.4% of ads. Men appeared alone but at work in 63.0% of ads. Men and women appeared together, outdoors, in 52.8% of ads. This is consistent with the studies conducted by Gilly (1988) and Jeryl and Jackson (1997).

Nevertheless, there is a statistically significant relationship between gender and *social role* ($p = 0.001 < 0.05$). Therefore, hypothesis 15 is supported. In 2007 (2017), in 72.2% (86.5%) of ads, women appeared alone in a modern role, while men appeared alone in 96.9% (90.7%) of ads. Women and men appeared together in a modern role in 92.5% (97.6%) of ads.

Concerning the religious allusion, there is a statistically significant relationship between gender and *religious allusion* ($p = 0.000 < 0.05$). Therefore, hypothesis 16 is supported. In 2007 (2017), women appeared in 11.1% (7.7%) ads containing religious innuendo, in 8.3% (5.8%) ads containing religious ceremonies, in 2.8% (40.4%) ads containing religious architecture, and also in 2.8% (17.3%) ads containing characters, 2.8% (32.7%) ads containing mythical places, and 2.8% (25%) in ads containing mythical objects. Men appeared alone in 6.3% (7.4%) ads containing religious ceremonies, 0.0% (11.1%) ads containing mythical objects, 3.1% (11.1%) ads containing religious innuendo. When they appeared together, men and women appeared in 14.3% (10.4%) ads containing religious innuendo.

Demographic Variables

Regarding demographic variables, we found a statistically significant relationship between gender and *age* for both men and women ($p = 0.000 < 0.05$): Therefore, hypothesis 4 is supported. In 2007 (2017), 44.4% (65.4%) of ads presented young women (under 35), and 25.0% (46.3%) presented men under 35. However, 45.9% (72.8%) of ads presented young women when the latter appeared with men, while 45.1% (70.4%) of ads presented men who were under 35 when they appeared with women. This agrees with the studies carried out by Gilly (1988), Wee et al. (1995a, 1995b, 1995c), Jeryl and Jackson (1997), Siu and Au (1997), and Millner and Higgs (2004).

When it comes to family status, we observed a statistically significant relationship between gender and *family status* for men as well as for women (p = 0.000 < 0.05). Therefore, hypothesis 5 is supported. As a matter of fact, in 2007 (2017), in 55.6% (90.4%) of ads there is no indication of the family status of the woman; 41.7% (9.6%) of ads presented married women and 2.8% (0.0%) single women. In contrast, 71.9% (98.1%) of ads showed no indication of the man's family status, 21.9% (0.0%) presented married men, and 6.3% (1.9%) single men. However, when they appeared with women in TV ads, men were married in 27.1% (4.8%) of ads, single in 21.1% (4.0%) of ads, and in 51.9% (91.2%) of ads there was no indication of their family status. As for women, when they appeared with men in TV ads, there was no indication of their family status in 49.6% (91.2%) of the ads; they were married in 35.3% (5.6%) and single in 15.0% (3.2%).

Concerning professional occupation, there is a statistically significant relationship between gender and the *job held*, for men as well as for women (p = 0.000 < 0.05). Therefore, hypothesis 6 is supported. In 2007 (2017), women appeared alone in 63.9% (65.4%) of ads, and men appeared alone in 71.9% (57.4%) of ads in which there was no indication of professional occupation. In 33.3% (32.7%) of ads, women appeared alone in a situation of unemployment, while they appeared alone but in a situation of employment in 2.8% (1.9%) of ads. Men, however, appeared alone in a working condition in 18.8% (40.7%) of ads and in a situation of unemployment in 9.4% (1.9%) of ads. Yet, when they appeared with women in TV ads, we observed that in 75.2% (79.2%) of the ads there was no indication concerning the men's jobs, and only in 14.3% (20.0%) of ads were men in a situation of work. On the other hand, men appeared as not working in 10.5% (0.8%) of ads, while there was no indication regarding work for women, who appeared with men in 72.9% (75.2%) of ads. As for women, 15.8% (20.8%) of ads featured them as not working, and 11.3% (4.0%) of ads showed them in a working situation. This is consistent with the studies made by Gilly (1988), Wee et al. (1995a, 1995b, 1995c), Jeryl and Jackson (1997), and Millner and Higgs (2004).

Moreover, there is a statistically significant relationship between gender and the *occupation* of men as well as of women (p = 0.000 < 0.05).

Therefore, hypothesis 7 is supported. In 2007 (2017), 63.9% (65.4%) of ads featured women and 71.9% (57.4%) featured men, with no indication about the occupation of any of them; 33.3% (28.8%) of ads presented housewives, while 21.9% (27.8%) of ads presented men having professional occupations. However, when they appeared with women in TV ads, there was no indication of men's occupation in 75.2% (79.2%) of cases. In 2007, 6.8% of ads men appeared as middle-level executives, while, in 2017, 9.6% of ads men appeared as student. Meanwhile, in 2007 (2017), when they appeared with men, women offered no indication of their occupation in 79.9% (75.2%) of ads. Only in 12.8% (9.6%) of cases did they appear as housewives. This is in keeping with the studies of Gilly (1988), Jeryl and Jackson (1997), and Millner and Higgs (2004).

Approach Variables

At the level of approach variables, we found a statistically significant relationship between gender and the *spokesperson*, both for men and for women ($p = 0.000 < 0.05$): Therefore, hypothesis 8 is supported. In 2007 (2017), 50.0% (32.7%) of ads showed women alone as spokespersons for the product or service; for men the figure is 65.6% (57.4%). However, 21.8% (32.0%) of ads featured women and men together as spokespersons for the product or service. This finding is consistent with the studies conducted by Jeryl and Jackson (1997) and Siu and Au (1997).

Hence, the relationship between gender and the *spokesperson's credibility* is statistically significant for men as well as for women ($p = 0.000 < 0.05$). Therefore, hypothesis 9 is supported. In 2007 (2017), 38.9% (32.7%) of ads showed women being spokespersons for a product when they were users of that product, whereas men were shown as spokespersons for a product when they were not users of it in 37.5% (35.2%) of ads. Nevertheless, when they appeared with women, we observed that in 18.0% (28.0%) of the ads, men were spokespersons when they were users of the product and only in 11.3% (8.8%) of the cases when they were not users. On the other hand, when they appeared with men, women were users of and spokespersons for a product in 21.1% (21.6%) of ads, and non-users of the product in 12.0% (11.2%) of ads. This agrees with the studies by Gilly (1988) and Jeryl and Jackson (1997).

As for the relationship between gender and *help*, there is a statistically significant relationship ($p = 0.000 < 0.05$). Therefore, hypothesis 10 is supported. In 2007 (2017), 97.2% (82.7%) of ads showed that women had no assistance. In 2.8% (15.4%) of ads, women appeared in a situation of offering help. Men on the other hand appeared as having no help in 87.5% (96.3%) of ads and as giving help in 6.3% (3.7%). This is consistent with the study by Millner and Higgs (2004).

Thus, there is a statistically significant relationship between gender and *advising* ($p = 0.000 < 0.05$). Therefore, hypothesis 11 is supported. Similarly, in 2007 (2017), 97.2% (82.7%) of ads showed that women are given some counseling, and 2.8% (15.4%) of ads showed them as providing counseling. Meanwhile, 81.3% (96.3%) of ads featured men as receiving no counseling and only in 18.8% (3.7%) of ads did they provide advice. This is in agreement with the study of Millner and Higgs (2004).

Moreover, there is a statistically significant relationship between gender and *role* ($p = 0.000 < 0.05$). Therefore, hypothesis 12 is supported. As a matter of fact, in 2007 (2017), 61.1% (96.2%) of ads showed that women were not dependent on others, which was the case for men in 84.4% (98.1%) of ads. Yet, when they were featured with women, men appeared as independent from others in 55.6% (91.2%) of ads, which was the case for women when they were featured with men in 53.4% (89.6%) of ads. This is consistent with the study of Millner and Higgs (2004).

Concerning the relationship between gender and *activity*, it is statistically significant ($p = 0.000 < 0.05$). Therefore, hypothesis 13 is supported. In 2007 (2017), women were shown as having no activity in 97.2% (98.1%) of ads, and men in 96.9% (77.8%) of ads. Thus, when women and men appeared together in advertisements, we observed that they had no activity. This is consistent with the studies of Siu and Au (1997) and Millner and Higgs (2004). Besides, there is a statistically significant relationship between gender and *frustration* for men as well as for women ($p = 0.000 < 0.05$). Therefore, hypothesis 14 is supported. In 2007 (2017), women appeared alone and having an unrelieved frustration in 97.2% of ads, and men in 93.8% of ads. In 2017, women and men appeared alone and having an unrelieved frustration in 100.0% of ads. This is consistent with the studies of Gilly (1988) in Mexican advertising.

Conclusions

The results of the content analysis of Tunisian TV ads in 2007 and 2017 suggest that most ads were related to services which were shot outdoors. Most characters were young who presented products intended for both men and women; their roles—which are modern—were not dependent on others, and there was no indication of their family status or their professional occupations. There was no help or counseling between men and women. Characters were inactive and had an unrelieved frustration. They were well dressed. In addition, in 2007, men's voices prevailed over women's voices. In 2017, women's voices prevailed over men's voices.

In cases where women appeared alone in TV ads, in 2007 and 2017, they presented products used by both men and women and products intended for women, and they took place at the household. In addition, the women were younger than the men and were not in a working situation, tended to be housewives. They were spokespersons for the products or the service when they were users of these products and were providers of help and advice. In addition, in 2007, women presented food products, while in 2017 they presented services.

In the case of men appearing alone in TV ads, in 2007 and 2017, the men presented services, products intended for both women and men, and masculine products. The featured men were younger than the women and were in a working situation, as executives. They were also providers of help and advice. In addition, in 2007, they took place outdoors and men were spokespersons for a product or service when they were not users of them, while in 2017, they took place at work and men were spokespersons for a product or service when they were users of them.

These results are inconsistent with the results of previous studies (Gilly, 1988; Wee et al., 1995a, 1995b, 1995c; Jeryl & Jackson, 1997; Siu & Au, 1997; Millner & Higgs, 2004), which reported that women appeared to present feminine products, while men more often presented masculine products. Men's voices were more common than women's. Women appeared at home, whereas men were seen at work or outdoors. The women also appeared younger than the men. There was no indication of family status for most men and women. Some of the characters appeared to be married. Women were often housewives, while men worked as executives. Women were spokespersons when they were users of the product, while men were spokespersons when they were not product users. Men and women appeared both as providers and as recipients of help and advice. They also appeared as inactive and having an unrelieved frustration.

When they were featured together in TV ads, in 2007 and 2017, men and women were shown outdoors, and they presented services and products intended for both of them. In addition, in 2007, women were featured as the receivers of help and counseling, whereas men were the providers of help and advice. In 2017, women and men were featured as the providers of help and advice.

The results indicate similarities between men and women when they are featured separately or together. In TV ads in 2007 and 2017, both women and men appeared inactive, performing modern roles in which they are independent from others, and having an unrelieved frustration.

The findings of the study have important implications for advertising practitioners. The results show that portrayals of men and women in Tunisian advertisements persist and are clearly becoming more prevalent since they reflect the reality and meet the requirements of the target receiver. This is consistent with the Cultivation Theory, which suggests that target viewers who are exposed to a particular view of the world through advertisements are more susceptible to advertising messages and the belief that they are real and valid (Fullerton & Kendrick, 2000; Gerbner, 1998). Thus, it is important that advertisers create advertising messages that reflect viewers' perceptions of social reality (A.E. Courtney & Whipple, 1983; Holbrook, 1987). As society's image toward men and women evolves, advertisements should represent a clean image in society and affirm it in the mind of the target receiver (Michell & Taylor, 1990).

Tunisia has deeply changed before the Tunisian revolution specifically during the month of Ramadan in 2007 and after the Tunisian revolution specifically during the month of Ramadan in 2017, and advertisements has changed with it. For example, in 2007 we found women represented in traditional role, while in 2017 women are represented in modern roles.

Tunisia has witnessed an increase in the number of working women with rising incomes. In fact, the woman was able to gain economic and financial independence because of her integration into the workforce. For this, women represented in advertising in traditional roles have decreased in Tunisia. Female characters are more often used in ads than male ones. This is explained by the fact that they have a more favorable impact in all fields. The image of women has undergone major evolutions in societies as well as in advertising. She passed from a traditional image which presented women as inferior to men, to modern image which reveals a degree of self-affirmation. This is why advertising executives now assign more importance to women in their ads. Several studies indicate that the use of women's images in advertising is growing.

In fact, women represent a more important target market. Indeed, advertisers have put more emphasis on the role of women in their advertising; they have to reflect their everyday lifestyle to be effective. Therefore, advertisers must represent characters that can be considered realistic (Wee et al., 1995a, 1995b, 1995c). An incorrect depiction of gender in advertising is harmful to society because it creates or perpetuates misconceptions. An unacceptable image portrayal in advertising could alter buying behavior (Lundstrom & Sciglimpaglia, 1977), and the use of stereotypical images discourages some consumers from buying products (Caballero et al., 1989; Callcott & Phillips, 1996; Jaffe & Berger, 1994).

Tunisian society has known cultural change in its predominant values; egalitarianism is a key element nowadays. Advertisements showing traditional women have decreased, and the representation of gender equality has increased. Thus, the depiction of women's sexism roles may irritate them as a consumer segment. Our results show that no advertisement has represented women at the level of sexism at the period analyzed after the Tunisian revolution specifically during the month of Ramadan in 2017, while 2.8% of female characters were classified at this level in 2007. The perception of women's and men's roles in advertising varies according to the culture of the country concerned. In Islamic culture, women are portrayed as occidental women. There is no nudity nor "hijab." Thus, we can expect to find differing representations of the roles of men and women in advertisements from different countries. Therefore, advertising practitioners should use appropriate portrayals of men and women to reflect Tunisian society values. Advertisers should break away from stereotypes and improve their marketing communications at the same time using dual roles, role switching, or role blending (Wee et al., 1995a, 1995b, 1995c). From the policy point of view, it is important that the Tunisian government issues regulations prohibiting gender-discriminatory advertising.

Limitations and Future Research

Although the current study provides advertisers with some important insights, it remains limited in scope. Neither the opinions of the creators and advertising editors, nor the consumers' perceptions were taken into account. Our study used a sample that is related to a specific period—the holy month of Ramadan in 2007 and 2017—and to a particular context—Tunisia. It would be judicious to extend the scope of this research by conducting a comparative study using similar countries (e.g., Morocco) or different countries (e.g., France) to understand the influence of social changes on advertising in Tunisia. The limitation of content analysis is that it cannot demonstrate the negative social effect of stereotyping and cannot analyze the effect of stereotyping on advertising effectiveness. Future researchers could examine whether ads with stereotype portrayals are more effective in generating sales or enhancing brand memory. Future research can explore consumers' perception about gender role portrayals in advertising and the consequences of the emotions generated by these cues on consumer attitudes toward advertisement.

Appendix 1

Definitions of measuring variables

Variables	*Authors*	*Descriptive criteria*
Product	Schneider and Schneider (1979)	What product category is presented in the ad?
Product user	Silverstein and Silverstein	The product featured in the ad targets: Women, men, both, or children.
Spoken comment	Dominick and Rauch (1972)	The announcing voice is feminine, masculine, both, no loud comment.
Shooting location	McArthur and Resko (1975)	The ad is shot at home, in a shop, outdoors, at work.
Social role of the character featured in the ad	*Sexton and Haberman in Marcelo et al.*	Traditional Modern
Age	Schneider and Schneider (1979)	The character's age is approximately -35, 35–50, 50+.
Family status	Schneider and Schneider (1979)	The featured character is married, single, status not identified.
Employment	Schneider and Schneider (1979)	The character appeared as a worker in a working situation, a worker in a non-working situation, no indication.
Occupation	Courtney and Lockeretz (1971)	The character appeared as a senior executive, a middle executive, a professional in the private sector, a student/pupil, at home.
Spokesperson	Schneider and Schneider (1979)	The product's on-camera spokesperson is a man, a woman, both, neither.
Credibility	McArthur and Resko (1975)	Characters were spokespersons when they were users of the product and non-users of the product.
Help	Silverstein and Silverstein	The character appeared as a help recipient, a help provider, neither.
Counseling	Silverstein and Silverstein	The character appeared as a receiver of advice, a provider of advice, neither.
Role	McArthur and Resko (1975)	The character's role is dependent on others, independent from others (spouse, sibling, housewife, worker, celebrity, interviewee, child, presenter).

(*continued*)

(continued)

Variables	*Authors*	*Descriptive criteria*
Activity	Poe	The character appeared as engaged in a physical activity/sport, active, or inactive.
Frustration	Silverstein and Silverstein	The character appeared as having a frustration that is thwarted or non-thwarted.

Sources: Wee et al. (1995a, 1995b, 1995c); Gilly (1988)

Appendix 2

Simple sorting: description of the ads investigated

Descriptive variables		*Percentages*	
		2007	*2017*
Product category	Services	45.3%	48.8%
	Food products	39.8%	28.2%
	Cosmetics	9.5%	2.1%
	Cleaning products	3.0%	6.9%
	Culture and entertainment	2.5%	14.1%
Shooting location	Outdoors	51.7%	38.1%
	At home	32.3%	23.4%
	At work	10.4%	19.6%
	In a shop	5.5%	3.8%
	No shooting location	0.0%	15.1%
Social role	Modern	89.6%	74.9%
	Traditional	10.4%	4.5%
	No social role	0.0%	20.6%
Product users	Both men and women	89.6%	94.5%
	Women	5.0%	1.7%
	Men	3.0%	2.7%
	Children	2.5%	1.0%
Spoken comments	Men	56.7%	30.2%
	Women	20.9%	40.2%
	Both men and women	11.9%	13.4%
	Neither men nor women	10.4%	16.2%

Appendix 3

Summary results of the crossed analysis (Chi-square and ANOVA) between descriptive variables and the presence of women and men or both

Descriptive variables		*2007*				*2017*			
		Presence of			*Sig. p*[a]	*Presence of*			*Sig. p*[a]
		Women alone	*Men alone*	*Both*		*Women alone*	*Men alone*	*Both*	
Product	Food products	61.1%	18.8%	39.1%	0.001	40.4%	31.5%	24.0%	0.000
category	Cleaning	0.0%	6.3%	3.0%		0.0%	1.9%	6.4%	
	products	22.2%	59.4%	48.1%		57.7%	44.4%	52.8%	
	Services	16.7%	6.3%	8.3%		1.9%	1.9%	0.0%	
	Cosmetics	0.0%	9.4%	1.5%		0.0%	20.4%	16.8%	
	Culture and								
	entertainment								
Shooting	At home	50.0%	21.9%	30.1%	0.005	40.4%	9.3%	32.0%	0.000
location	In a shop	5.6%	3.1%	6.0%		7.7%	1.9%	4.0%	
	Outdoors	41.7%	65.6%	51.1%		38.5%	25.9%	52.8%	
	At work	2.8%	9.4%	12.8%		13.5%	63.0%	11.2%	
Social role	Traditional	27.8%	3.1%	7.5%	0.001	13.5%	9.3%	0.8%	0.000
	Modern	72.2%	96.9%	92.5%		86.5%	90.7%	97.6%	
Product	Women	16.7%	0.0%	3.0%	0.000	9.6%	0.0%	0.0%	0.000
user	Men	0.0%	12.5%	1.5%		1.9%	7.4%	0.0%	
	Both	77.8%	87.5%	93.2%		88.5%	92.6%	97.6%	
	Children	5.6%	0.0%	2.3%		0.0%	0.0%	2.4%	
Spoken	Women	47.2%	6.3%	17.3%	0.000	59.6%	5.6%	47.2%	0.000
comments	Men	30.6%	81.3%	57.9%		9.6%	40.7%	26.4%	
	Both	13.9%	3.1%	13.5%		11.5%	5.6%	21.6%	
	Neither	8.3%	9.4%	11.3%		19.2%	48.1%	4.8%	
Religious	Religious	2.8%	0.0%	7.5%	0.000	40.4%	9.3%	6.4%	0.000
allusion	architecture	8.3%	6.3%	4.5%		5.8%	7.4%	9.6%	
	Religious	2.8%	0.0%	5.3%		17.3%	3.7%	11.2%	
	ceremonies	2.8%	0.0%	10.5%		32.7%	3.7%	9.6%	
	Characters	2.8%	0.0%	9.0%		25.0%	11.1%	11.7%	
	Mythical	11.1%	3.1%	14.3%		7.7%	11.1%	10.4%	
	places								
	Mythical								
	objects								
	Religious								
	innuendo								

[a]$p<0.05$

Appendix 4

Summary results of the crossed analysis (Chi-square) between demographic and approach variables and the presence of men, women, or both

Variables		*2007*				*2017*			
		Presence alone of		*Presence both of*		*Presence alone of*		*Presence both of*	
		Women	*Men*	*Men*	*Women*	*Women*	*Men*	*Men*	*Women*
Age	Under 35	44.4%	25.0%	45.1%	45.9%	65.4%	46.3%	70.4%	72.8%
	30–35	27.8%	37.5%	30.1%	36.1%	15.4%	16.7%	16.8%	19.2%
	Over 35	27.9%	37.5%	24.8%	18.0%	19.2%	37.0%	12.8%	8.0%
Family status	Married	41.7%	21.9%	27.1%	35.3%	9.6%	0.0%	4.8%	5.6%
	Single	2.8%	6.3%	21.1%	15.0%	0.0%	1.9%	4.0%	3.2%
	Not identified	55.6%	71.9%	51.9%	49.6%	90.4%	98.1%	91.2%	91.2%
Employment	Working	2.8%	18.8%	14.3%	11.3%	1.9%	40.7%	20.0%	4.0%
	Not working	33.3%	9.4%	10.5%	15.8%	32.7%	1.9%	0.8%	20.8%
	Not mentioned	63.9%	71.9%	75.2%	72.9%	65.4%	57.4%	79.2%	75.2%
Occupation	Senior executive	0.0%	0.0%	3.0%	0.8%	1.9%	5.6%	0.8%	1.6%
	Middle executive	0.0%	0.0%	6.8%	8.3%	0.0%	5.6%	0.0%	0.0%
	Professional job	2.8%	21.9%	6.0%	2.3%	3.8%	27.8%	9.6%	7.2%
	Student/pupil	0.0%	0.0%	4.5%	3.0%	0.0%	1.9%	9.6%	6.4%
	At home	33.3%	6.3%	4.5%	12.8%	28.8%	1.9%	0.8%	9.6%
	Not mentioned	63.9%	71.9%	75.2%	79.9%	65.4%	57.4%	79.2%	75.2%
Spokesperson	Woman	50.0%	0.0%	10.5%		32.7%	0.0%	0.8%	
	Man	0.0%	65.6%	7.5%		0.0%	57.4%	5.6%	
	Both	0.0%	0.0%	21.8%		0.0%	0.0%	32.0%	
	Neither	50.0%	34.4%	60.2%		67.3%	42.6%	61.6%	

Spokesperson's credibility	As product user	38.9%	2814%	18.0%	21.1%	32.7%	35.2%	28.0%	21.6%
	As a non-user	11.1%	37.5%	11.3%	12.0%	0.0%	22.2%	8.8%	11.2%
	No spokesperson credibility	50.0%	34.4%	70.7%	66.9%	67.3%	42.6%	63.2%	67.2%
Help	Help recipient	0.0%	6.3%	2.3%	4.5%	1.9%	0.0%	7.2%	4.8%
	Help provider	2.8%	6.3%	5.3%	2.3%	15.4%	3.7%	9.6%	11.2%
	No help	97.2%	87.5%	92.5%	93.2%	82.7%	96.3%	83.2%	83.2%
Counseling	Advice recipient	0.0%	0.0%	3.8%	5.3%	1.9%	0.0%	7.2%	4.8%
	Advice provider	2.8%	18.8%	6.0%	3.8%	15.4%	3.7%	9.6%	11.2%
	No counseling	97.2%	81.3%	90.2%	91%	82.7%	96.3%	83.2%	83.2%
Role	Dependent on others	38.9%	15.6%	44.4%	46.6%	3.8%	1.9%	8.8%	9.6%
	Independent from others	61.1%	84.4%	55.6%	53.4%	96.2%	98.1%	91.2%	89.6%
Role	Spouse	13.9%	6.3%	13.5%	15.0%	0.0%	1.9%	3.2%	3.2%
	Sibling	16.7%	3.1%	10.5%	18.8%	3.8%	0.0%	14.4%	9.6%
	Housewife	8.3%	3.1%	0.0%	3.0%	19.2%	1.9%	1.6%	5.6%
	Worker	0.0%	21.9%	13.5%	9.8%	0.0%	9.3%	4.8%	2.4%
	Celebrity	27.8%	37.5%	6.0%	4.5%	32.7%	25.9%	15.2%	15.2%
	Interviewee	0.0%	0.0%	0.8%	0.0%	1.9%	0.0%	0.0%	0.0%
	Child	2.8%	3.1%	18.8%	12.0%	0.0%	1.9%	12.0%	11.2%
	Presenter	30.6%	25.0%	36.8%	36.8%	42.3%	59.3%	48.8%	52.0%
Activity	Active	2.8%	3.1%	2.3%	1.5%	1.9%	22.2%	14.4%	14.4%
	Inactive	97.2%	96.9%	97.7%	98.5%	98.1%	77.8%	85.6%	84.8%
Frustration	Thwarted	2.8%	6.3%	1.5%	1.5%	0.0%	0.0%	0.0%	0.0%
	Non-thwarted	97.2%	93.8%	98.5%	98.5%	100.0%	100.0%	100.0%	100.0%

Note

1. Source: document «Mediascene», «*Medmedia*» (Agence media) and «*Mediascan*» (institut of Mediametrie)

References

Belch, M., Belch, G., Ballofet, P., & Coderre, F. (2004). *Communication Marketing: Une Perspective Intégrée*. McGraw Hill.

Caballero, M. J., Lumpkin, J. R., & Madden, C. S. (1989). Using Physical Attractiveness as an Advertising Tool: An empirical Test of the Attraction Phenomenon. *Journal of Advertising Research, 27*, 16–22.

Callcott, M. F., & Phillips, B. J. (1996). Elves Make Good Cookies: Creating Likable Spokescharacter Advertising. *Journal of Advertising Research, 36*, 73–79.

Courtney, A. E., & Lockeretz, S. W. (1971, February). A Women's Place: An Analysis of the Role Portrayed by Women in Magazine Advertisements. *Journal of Marketing Research, 8*, 92–95.

Courtney, A. E., & Whipple, T. W. (1983). *Sex Stereotyping in Advertising*. Lexington Books.

Das, M. (2000). Men and Women in Indian Magazine Advertisements: A Preliminary Report. *Sex Roles, 43*, 699–717.

Décaudin, J.-M. (2003). *La Communication Marketing, Concept, Techniques, Stratégies* (3ème ed.). Économica.

Dominick, J. R., & Rauch, G. E. (1972, Summer). The Image of Women in Network TV Commercials. *Journal of Broadcasting, 19*, 265–259.

Dupond, L. (1993). *1001 trucs publicitaires* (2 ème ed.). Transcontinentale.

Ferguson, J. H., Kreshel, P. J., & Tinkham, S. F. (1990). In the Pages of Ms.: Sex Role Portrayals of Women in Advertising. *Journal of Advertising, 19*(1), 40–51.

Ford, J., Latour, M., Honeycutt, D., & Joseph, M. (1994, May). Female Sex Role Portrayals in International Advertising: Should Advertisers Standardise in the Pacific Rim. *American Business Review, 12*, 1–10.

Ford, J., Latour, M., & Miileton, C. (1999). Women's Studies and Advertising Role Portrayal Sensitivity: How Easy Is It to Raise Feminist Consciousness. *Journal of Current Issues and Research in Advertising, 2*(2), 77–87.

Ford, J., Latur, M., & Lundstrom, W. (1991). Contemporary Women's Evaluation of Female Role Portrayals in Advertising. *The Journal of Consumer Marketing, 8*(1), 15–28.

Fullerton, J. A., & Kendrick, A. (2000). Portrayal of Men and Women in US Spanish Language Television Commercials. *Journalism and Mass Communication Quarterly, 77*(1), 128–142.

Gavard-Perret, M. L. (1993). La présence humaine dans l'image, facteur d'efficacité de la communication publicitaire: Une expérimentation dans le domaine du tourisme. *Recherche et Application Marketing, 8*(2), 1–22.

Gerbner, G. (1998). Cultivation Analysis: An Overview. *Mass Communication and Society, 1*(Summer), 175–194.

Gilly, C. M. (1988, April). Sex Roles in Advertising: A Comparison of Television Advertisements in Australia, Mexico, and the United States. *Journal of Marketing Research, 52*(2), 75–85.

Holbrook, M. B. (1987). Mirror, Mirror on the Wall, What's Unfair in the Reflections on Advertising. *Journal of Marketing, 51*(July), 95–103.

Jaffe, L. J., & Berger, P. D. (1994). The Effect of Modern Female Sex Role Portrayals on Advertising Effectiveness. *Journal of Advertising Research, 32*, 32–42.

Jeryl, W., & Jackson, D. (1997). Women in TV Advertising: A Comparison between the UK and France. *European Business Review, Bradford, 97*(6), 294.

Klassen, M. L., Jasper, C. R., & Schwartz, A. M. (1993, March–April). Men and Women: Images of Their Relationship in Magazine Advertisements. *Journal of Advertising Research, 33*, 30–39.

Lerman, D., & Callow, M. (2004). Content Analysis in Cross-Cultural Advertising Research: Insightful or Superficial. *International Journal of Advertising, 23*(4), 507–521.

Lo'pez Pintor, R. (1990). *Sociologı'a Industrial.* Alianza Editorial.

Lundstrom, W. J., & Sciglimpaglia, D. (1977). Sex Role Portrayals in Advertising. *Journal of Marketing, 41*(3), 72–79.

Manstead, A. S. R., & Culloch, M. C. (1981). Sex role Stereotyping in British Television Advertisements. *British Journal of Social Psychology, 20*, 171–180.

Mays, A. E., & Brady, D. L. (1991). Women's Changing Role Portrayals in Magazine Advertisements: 1955–1985. *Journal of Advertising, 20.*

McArthur, L. Z., & Resko, B. G. (1975, December). The Portrayal of Men and Women in American Television Commercials. *Journal of Social Psychology, 97*, 209–220.

Michell, P. C. N., & Taylor, W. (1990). Polarising Trends in Female Role Portrayals in UK Advertising. *European Journal of Marketing, 24*(5), 41–49.

Millner, L. M., & Higgs, B. (2004). Gender Sex Role Portrayals in International Television Advertising Over Time: The Australian Experience. *Journal of Current Issues and Research in Advertising, 26*(2), 81–95.

O'Donnell, W. J., & O'Donnell, K. J. (1978). Update: Sex-Role Messages in TV Commercial. *Journal of Communication, Winter*, 156–158.

Pollay, R. W. (1986, April). The Distorted Mirror: Reflections on the Unintended Consequences of Advertising. *Journal of Marketing, 50*, 18–38.

Resnik, A., & Stern, B. L. (1977). An Analysis of Information Content in Television Advertising. *Journal of Marketing Research, 41*, 50–53.

Schneider, K. C., & Schneider, S. B. (1979). Trends in Sex Role Television Commercials. *Journal of Marketing Research, Summer, 43*, 79–84.

Shrum, L. J., Wyer, R. S., & O'Guinn, T. C. (1998, March). The Effects of Television Consumption on Social Perceptions: The Use of Priming Procedures to Investigate Psychological Processes. *Journal of Consumer Research, 24*, 447–458.

Siu, W. S., & Au, A. M. (1997). Women in Advertising: A Comparison of Television Advertisements in China and Singapore. *Marketing Intelligence and Planning, Bradford, 15*(5), 235.

Taylor, C. R., Miracle, G. E., & Wilson, R. D. (1997). The Impact of Information Level on the Effectiveness of U.S. and Korean Television Commercials. *Journal of Advertising, 26*(1), 1–15.

Walliser, B., & Moreau, F. (2000). Comparaison du style français et allemand de la publicité télévisée. *Décision Marketing, Janvier - Avril, 19*, 75–84.

Wee, C.-H., Choong, M.-L., & Tambyah, S.-K. (1995a). Sex Role Portrayal in Television Advertising. *International Marketing Review, 12*(1), 49.

Wee, C., Choong, M., & Tambyah, S. (1995b). Sex role Portrayal in Television Advertising a Comparative of Singapore and Malaysia. *International Marketing Review, 12*(1), 49.

Wee, C. H., Choong, M. L., & Tambyah, K. K. (1995c). Sex Role Portrayal in Television Advertising: A Comparative Study of Singapore and Malaysia. *International Marketing Review, 12*(1), 49–64.

Wiles, J. A., Wiles, C. R., & Tjernlund, A. (1995). A Comparison of Gender Role Portrayals in Magazine Advertising: The Netherlands, Sweden and USA. *European Journal of Marketing., 29*(11), 35–49.

Zhou, N., & Chen, M. (1997). A Content Analysis of Men and Women in Canadian Consumer Magazine Advertising: Today's Portrayal, Yesterday's Image. *Journal of Business Ethics, 16*(5), 485.

CHAPTER 3

Women, Financial Inclusion, and Economic Development in Rwanda

Madelin O'Toole

Introduction

Economic development depends on numerous factors that vary from nation to nation. Cultural, institutional, geographic, and historical components all play a role in a country's development path. Over the last few decades the focus of development research has shifted from general plans that were applied to all less developed countries (LDCs) to understanding that development is different for every nation; therefore, research must focus on individual characteristics. Methods of development have evolved along with this new understanding; areas like finance, gender equality, small business development, and improvements to infrastructure are all becoming preferred routes of increasing development. Many NGOs and IGOs now focus on financial inclusion and gender equality. These variables are of great importance individually but are even more so when considered in unison. Development must be inclusive to continue in the long

M. O'Toole (✉)
Federal Research Division, Library of Congress, Washington, DC, USA

B. S. Nayak (ed.), *Political Economy of Gender and Development in Africa*, https://doi.org/10.1007/978-3-031-18829-9_3

term; financial access and gender equality are methods to ensure long-lasting stability.

The development of financial markets is both a result and driver of economic development. According to Arestis and Dimitriades (1997) the direction of the relationship cannot be declared absolutely, but instead must be observed as mutual influencers on one another's growth. Financial services act as an indicator to the level of development for all nations, where the most developed have an abundance of financial services with the sector making up a large component of the economy. The financial sector acts as an indicator of growth for less developed countries, as well, where increased account numbers and micro-loans indicate growth in wages and business start-ups. The presence of well-constructed financial markets ranges from the smallest deposit account all the way to a complex stock market system that allows for individuals to invest and earn returns. In the case of many LDCs the data on such factors is scarce, but recent initiatives by the World Bank and IMF have made the collection and analysis of such data paramount to development plans. The growing concern of financial inclusion is well justified, however, much of the policies intended to increase access could very well lead to more exclusion. Manji (2010) notes that the campaign to allow land titles as collateral for loans inadvertently creates an inequality in financial access between genders. For this reason, the issue of financial institutions must also be analysed directly with gender inequality. Should nations accept policy changes that create greater access for some but bar others from participating, the path to growth will become much more difficult.

Economic equality between genders has increased over time, however, a disparity between genders still exists. In the world's most developed nations women still face inequality on the basis of gender. The numbers prove women still face a gap in wages, with that gap increasing over a woman's career. It is also found that nations with a smaller wage gap usually have lower female participation in the labour force; and if the difference in participation based on gender is taken into account the wage gap becomes even larger. Women are also disproportionally overrepresented in low paying jobs, meaning that the higher rates of employment amongst women are typically in low paying sectors. There has been an increase of women in higher paying jobs over time, but these positions are still male dominated (Issac, 2014). Employment and wages are just one component of economic inequality between genders.

In many cases control over household earnings remains with the male head of the household, with a number of countries dictating this in their law system. The lack of autonomy in deciding how to spend personal income is especially common in low-income regions like sub-Saharan Africa, Asia, and the Middle East (Issac, 2014). The lack of autonomy that extends beyond financial decisions and into other areas of daily life is significant, leaving women to bear the burden of another's decisions. The lack of female input and control of finances further deepen the origins of inequality in developing nations. Women in more developed nations are better able to access financial services, make decisions for the household, and choose how to spend their income. The freedom for females to access financial markets and services is vital to closing the equality gap in developing nations. In Pakistan, women are able to open bank accounts but the decision making for the money in those accounts still remains with a male family member. Other nations require the permission of a male relative just to open the account which severely restricts personal and financial freedom. Just recently women in Côte d'Ivoire can now legally be the head of a household meaning that they are able to claim tax deductions that result increased financial independence (Issac, 2014). Some of the cases mentioned previously are rare, but inequalities still exist that put women at a severe disadvantage. The lack of access to general financial services like savings and debit accounts is just one component of inequality found in financial institutions. Women starting businesses often find it increasingly difficult to get funding for such ventures. According to a report by John Isaac (2014), 30%–37% of all SMEs in emerging economies are female owned; however, female founded businesses have nearly $300 billion in unmet funding (Issac, 2014). Access to funding is just another barrier of entry met by businesses in emerging markets, thus making it difficult for nations to develop to their full potential.

The lack of female participation in a key area of the economy is not a solely female issue. The barriers in accessing financial services affect the families of these women as well and lead to residual effects in the rest of the economy. The lack of loans for starting businesses and access to accounts to keep wages secure reduce productivity of the entire economy. According to Ospina and Roser's report on economic inequality by gender, women are less likely to borrow capital for productive purposes across all countries studied. Economic growth is reliant on the active participation of women in the economy, especially in developing nations. Without full and equal access to financial services for all nations, they will never reach their full economic potential.

The importance of financial inclusion and gender equality to development is paramount. The relationships must be understood for future policy making to be effective in developing nations. In an effort to understand the effects of gender equality and access to finance, the East African country of Rwanda is analysed. After the 1994 genocide the nation faced a depleted population with a female majority. In order to rebuild and move on from the trauma of the mass killings an effort was made to ensure the equal inclusion of women in government, education, and the economy. Since Rwanda is rated as one of the most equal in the world, with more than half of the parliament being female, the relationship between equality, finance, and development provides a unique case study. Therefore, the question that is to be answered is: What is the influence of financial inclusion and gender equality on Rwanda's development? To answer this question a literature review analyses key papers involving development, access to finance, and finally the role of gender equality in both. Key empirical findings and theory are utilised in detail for the construction of the models that are tested. Four time-series regressions determine the relationship and magnitude of selected development, finance, and equality variables. Finally, the results determine the effects of gender equality and finance on Rwanda's development over the past decade.

Financial Access and Development

The current literature surrounding development, financial inclusion, and gender equality suggests that the three must occur simultaneously. The first section of this chapter shows the connections between development and access to finance by explaining the findings of two highly cited papers by Arestis and Dimitriades (1997) and Sarma et al. The authors come to the conclusion that well-developed and efficient financial sectors are key indicators of developed nations, but for development to last social institutions must be in place to ensure gender equality and property rights. The second section of the literature review analyses the literature on inequality and access to finance. The authors in this section find a strong relationship between inequality measures, like the Gini coefficient (Ang, 2010; Batuo et al., 2010), and accessibility to finance for numerous developing nations. Findings from this section suggest that decreased equality increases financial inclusion, and increased inclusion decreases inequality. Section three adds the component of gender equality to the mix, with authors like Kabeer (2005), van Staveren (2001), and Klapper and Parker (2011) finding that females both on the firm and household levels face inequality when accessing finance. Section four dives deeper into the issue of gender inequality and access to finance by analysing the relationship between developing nations. All authors in this section come to the conclusion that exclusion is not a result of discrimination by the financial institution, but instead results from institutional inequalities. Section four furthers the support of gender inequality's role in slow development and limited financial inclusion, especially in Africa. Hansen and Rand (2014) and Johnson and Nino-Zarazua (2011) conclude that any inequality created by social institutions has a significant negative impact on inclusion and development. The evidence for the mutual relationship between equality, inclusion, and growth is well supported throughout development literature and empirical findings.

Development, Inequality, and Access to Financial Services

This section outlines development and access to finance without the gender dimension presented as it provides valuable methodology and findings that can be applied to the construction of research focused on gender inequality and access. The next section focuses primarily on the most recent findings on gender inequality and access to financial services across developing nations and Sub-Saharan Africa. Much of the methodology and reasoning from the literature form the basis of the research methods to be utilised in the case of Rwanda. The gaps in the literature are pointed out as well, so that future research on the topic can address any weak or missing components.

Arestis and Dimitriades (1997) raise valuable points in regard to financial development and its role in overall development of poorer nations. First, the link between finance and development is heavily dependent on the characteristics of the country's institutions. Second, the direction of the relationship between finance and development cannot be defined one way or the other. Both development and finance increase as the other increases and which variable causes changes in the other is dependent on the individual nation. The authors, therefore, stress the importance of analysing each nation on an individual basis as opposed to cross-country analysis as it captures the circumstances that influence the financial system of each nation. The results of their efforts do not allow for conclusive results that are applicable to all nations. Instead, the authors provide guidance for future studies, stressing the importance of utilising time-series data for individual countries. By designing future studies in this way, the individual country characteristics can be captured, and the results useful.

Sarma et al. investigate the relationship of financial inclusion and development by conducting a cross-country analysis. The authors discover that financial exclusion is a result of social exclusion. Development and financial inclusion prove to be positively correlated over the cross-country analysis, with social factors also playing a key role in both access and development. Inequality especially makes or breaks the development path of LDCs proving that unequal societies will always be limited in development. While, the authors focus mainly on the general inequality of income and other social components, gender equality is not specifically addressed; however, the same logic and methods apply just as well to gender inequality. Socio-cultural and institutional factors severely influence development and when combined with the component of financial access prove that development must be an inclusive process.

Inequality and Access to Finance in Developing Nations

General economic inequality stems from inefficiencies in financial access. In an article for the World Bank, authors Beck et al. (2009) found that the lack of access not only drives economic inequality but also hinders future growth. The authors argue that government policies are key to closing the gap in equality. They suggest that infrastructure, protection of property rights, and competition are essential changes in policy (Beck et al., 2009). While this article does provide insights to important policy requirements for solving general inequality, it does not specifically address the gender equality component. The disparity between genders must also be included and policies must be tailored to combating economic gender inequality. The authors highlight the importance of property rights and infrastructure, both critical components for ensuring equal access and participation for females. Institutions like education, employment opportunity, and property rights all create a level playing field between genders. If the same principles are tailored to include more women, the overall impact of financial services on the entire economy would increase significantly.

James Ang (2010) investigates the relationship between finance and income inequality in India over fifty years. The main goal of the research is to specifically analyse the relationship between financial development and liberation on the Gini coefficient. Underdeveloped financial systems and in turn services have a larger negative impact on the poor increasing income inequality. The presence of well-developed and efficient financial services contributes to increased equality, growth, and development. While this research only focuses on general income inequality and access to financial services, it does provide some insight on the subject of gender inequality. Women in developing nations already face lower wages or remain unemployed entirely, therefore impacting them more severely than the population as a whole. If income inequality decreases as access to financial services increases, then the inequality between genders would also decrease. However, if this access remains unequal between genders and not just between income levels the gap would widen. Ang's (2010) research lacks detailed analysis on the effect of gender inequality but does provide a basis for this to be investigated in the future. Depending on specific national policy implications the gender gap could be significantly impacted by wider and efficient access to financial services.

In an analysis focusing on 37 Asian Pacific nations, Park and Mercado (2015) investigate the relationship between financial inclusion and income inequality. Their findings support the importance of per capita income, demographic factors, and institutional factors. It also proves a significant and robust negative correlation between poverty and income inequality and increased financial access (Park & Mercado, 2015). There was no additional research on the gender component of inequality, but it was found that demographic characteristics like large population size and low dependency ratios are associated with higher access to financial services. These findings provide evidence that higher financial inclusion for women would be an additional method for decreasing inequality between genders.

Batuo et al. (2010) focus their research on financial development and income inequality in Africa. The authors test the relationship between the measure of inequality, the Gini coefficient, and financial development to determine if there is a linear or inverted U shape in the results as suggested by previous literature. They find a negative linear relationship between finance and Gini coefficients, meaning that as access to financial services increases, income inequality decreases. There is no U shape detected, therefore, it is assumed financial development continuously reduces income inequality over time (Batuo et al., 2010). Their findings suggest that it is vital for African nations to encourage access to financial services in order to close the gap in inequality. Rural populations and women would benefit especially from an increase as they are the most affected by lack of access.

Gender Inequality and Access to Finance

Access to financial services is just one aspect of closing the inequality gap but direct financing through microloans is growing in popularity. Prof. Naila Kabeer focuses on the use of microfinance for the empowerment of women. She defines access to financial services as a method of closing the gender gap and suggests that microfinance as a more direct method of providing access to financial services. She notes that financial systems focus on individualism and are better able to "deal with the financial exclusion of the poor and market failures" (Kabeer, 2005). Unequal access and participation to financial services are a result of entrenched social constraints, such as lack of education, employment opportunities, and ethnic and gender exclusions (Kabeer, 2005). Access to financial markets is a result of institutional discrimination and not a failure of financial services themselves. The degree to which these components directly affect unequal access is not directly addressed by Kabeer (2005) but leaves a gap in the literature for future statistical analysis.

In a report of World Bank, John Issac (2014) stated "Women disproportionately face barriers to accessing finance that prevent them from participating in the economy and improving their lives" (Issac, 2014). Female-owned businesses make up around 40% of small to medium firms in developing markets; however, a majority of the financial needs are unmet by current financial services. This is partly attributed to the industry in which these businesses operate, but government and cultural restrictions are also to blame for poor access. A lack of data and capacity make solving this issue increasingly difficult, as most data on household behaviours is self-reported through surveys making the accuracy less certain.

In an article for *Gender and Development*, van Staveren (2001) analyses gender bias in finance. The author notes that at the micro-level men and women are comparatively different on financial behaviour. The difference in behaviour can be attributed to several factors, but one that is noted in particular are differences in income. In developing countries women participate in unpaid work, such as household care and child rearing, or are employed in low-skill and low-wage work. The initial factors of employment and pay give an unequal basis for financial inequality to grow further. van Staveren (2001) finds "gender distortions" in financial markets, which create a disadvantage for females as they also lack collateral to participate in loan activity. Financial illiteracy actively discourages women from accessing services, especially in countries with low general literacy rates. Overall, men have a better understanding of financial products and services since they are more likely to access financial services. When women are able to invest and participate, they are actually better investors and have higher repayment rates on loans than men, assuming the woman retains control after receiving the loan. However, the author notes that in a study conducted on the Grameen Bank only 37% of female borrowers retain control of their loans (van Staveren 2001). Even if women are able to access loans, control after borrowing is another indicator of recurring inequality.

Klapper and Parker (2011) conduct a literature review for the *World Bank Observer* titled *Gender and the Business Environment for New Firm Concentration*. The authors summarise the differences between male- and female-owned firms. Firms owned by men and women typically are divided not only by gender but by industry as well. Female entrepreneurs are mainly concentrated in labour-intensive, small, and highly competitive sectors. Businesses in this sector rely heavily on informal financial services and require less funding than a capital-intensive high-skilled operation, especially for developing countries. Barriers in accessing financial services could also lead women to take up business in high labour sectors as opposed to capital intensive. Klapper and Parker (2011) note that in African and Middle Eastern nations, women must have the permission of a male family member to apply for loans, creating a higher barrier of entry. Lack of education, initial funds, and business experience further increase barriers that make the process of obtaining business loans increasingly difficult (Klapper & Parker, 2011). From the literature reviewed, there is no explicit discrimination from the financial institutions found, but most barriers are created by institutional and social factors of the nation. There is some evidence that women did face higher interest rates for the loans, but other factors like less start-up capital provided by the entrepreneur could account for the higher interest rate. When women do receive loans for business there is a positive effect on economic growth in the nation.

Access to Finance in Developing Nations

In the works of Chavan (2008), "extent and nature of gender inequality in the provision of banking services in India" are clear (Chavan, 2008). The author uses two indicators to represent access to banking services: level of credit supplied to women and level of deposits received by women. Chavan (2008) breaks down the data into subgroups to compare access between rural and urban areas; a large disparity between the two is found. Women in urban areas have significantly better access to financial services than women in rural areas. Geographic location plays a vital role in a woman's ability to access financial services. Travel time, safety, and potential loss of productivity discourage women from accessing financial services, especially in rural areas. The study also found that proportionately, women contribute more in the deposits than they receive credit loans than men from financial institutions. Females can access financial services but are still limited in participation, highlighting the key issue of inequality between genders. There are a few reasons that men take out more loans than females: men are more willing to take risk, men make the decisions for the whole household, and men are more likely to take out loans to start businesses. Overall, Chavan (2008) points out key factors that contribute to unequal access to financial services between genders, but within females as well.

Unequal access to financial services and limited participation stem from deeper cultural practices that limit female autonomy to actively utilise services. According to Seema Jayachandran (2015), there is more gender inequality in poor developing nations than in wealthier developed nations. The author attributes the causes of the inequality to a lack of autonomy and participation in the workforce, and therefore the ability to earn an income and make financial decisions. In poor nations, access to education for girls is much less valued than it is for boys. The lack of access to education combined with the fact that women are already undervalued in society create a cyclical effect that leads to more unemployed women, thus, continuing the inability to earn an income that would allow for participation in financial services in the first place. In Northern Africa, the Middle East, and India women's mobility is also restricted. This deepens the previously mentioned causes of inequality as women lack freedom of movement. Consistently, traditional gender roles perpetuate the unequal access to education, employment, and access to financial services. It is important to include the causes of inequality when analysing unequal access to financial services in developing nations, as the specific country characteristics drive initial inequality that leads to inequality in accessing financial services.

Access to Finance in Africa

Blackden et al. research the disparity between gender inequality and growth in Sub-Saharan Africa. The authors note gender-based barriers impede growth in the region, where poverty is the highest among developing nations. Gender inequality spans multiple dimensions but education, employment, control of assets, and governance issues are the major areas highlighted by Blackden et al. All areas tested are significant in contributing to gender inequality for the region; however, it is important to note that participation and access to financial services are not addressed by the authors. There are linkages between education, employment, and accessing financial services and usually the higher the education and employment levels the greater the usage of financial services. Growth in Sub-Saharan Africa has fallen behind other developing nations because of the underutilisation of women. If given equal education, employment, freedoms, and access to financial services growth in Sub-Saharan Africa would at the very least catch up and keep pace with the rest of the developing world.

Manji (2010) analyses development policy aimed at increasing participation in financial services for Eastern Africa and the effect the policies have on gender inequality in the region. The author notes that previous World Bank reports emphasise reforming the financial sector to promote growth and access to financial services. The use of land titles as collateral for accessing credit and participating in financial services is one main suggestion by the World Bank; however, Manji (2010) points out some key issues with this policy regarding equality. Land ownership is predominantly male with many nations restricting ownership by law; therefore, females are already at a disadvantage as they are unlikely to own land. If World Bank policy is focused solely on increasing access to financial services by exclusively using land as a qualifier, then all non-land owners and nearly all women will be excluded from participating. Thus, usage of this policy would continue a pattern of gender inequality in the region as well as increase the negative long-term effects of limited access to financial services. When half of the population is excluded from participating in key part of the economy the entire nation suffers. Manji (2010) argues that this World Bank policy is inherently biased and only perpetuates the cycle of inequality that exists in cultural practices. The author raises a major issue with policy put forth by supranational agencies like the World Bank, many policies that are meant to help inadvertently create more inequality, especially gender inequality, and hurt development in the long run. Therefore, it is important to keep this in mind and remain critical when analysing policy and inequality.

Aterido et al. (2013) break down their analysis between equality of financial services between personal accounts and business accounts. Access to finance for business includes formal and informal financing and the difference between genders in accessing formal financing channels. Initially findings suggest that there is a gender gap however, once controlled for external characteristics the authors find no difference between male and female access of formal financing for business (Aterido et al., 2013). Firm characteristics appear to create the inequalities present, especially size. Selection bias contributes to observed inequality, as female entrepreneurs have higher barriers of entry than their male counterparts. There is no inequality between genders at the household level in use of formal services once controlled for other characteristics. The main finding of this paper suggests that inequality is not created by financial services themselves but instead external characteristics that influence ability and willingness to access formal financial services. Education, income level, employment, and age are all key characteristics that affect accessing formal financial services.

Asiedu et al. (2003) focus on access to financing for female firms in Sub-Saharan Africa. The authors note that current empirical findings suggest lack of access to financing is the main constraint to all firm growth (Asiedu et al. 2003). Due to the lack of economic stability in Africa the private sector remains underdeveloped. The constraints faced by African firms are even more challenging for female-owned firms; however, the key to growth and a stronger private sector depends on female involvement. The authors find that female firms in general face more constraints than male-owned firms. Sub-Saharan firms are more constrained than other regions and these constraints significantly impact females in the region. Asiedu et al. (2003) found that female-owned firms are constrained by 5.2 percentage points more than male-owned firms in the area. Even after controlling for other characteristics and bias the gender gap persisted, highlighting that females are disadvantaged when acquiring financial support for business purposes.

Hansen and Rand (2014) test the accuracy of measures used to determine inequality of firm's credit access in Sub-Saharan Africa based on gender. The authors test the accuracy of self-reported survey data versus formal financial data. Self-reported data measures the perceived inequality between male- and female-owned businesses in accessing credit. When using this measure the authors find a "marginally significant gender difference for the credit perception" where female firms are more constrained by lack of access to financing than male firms (Hansen & Rand, 2014). However, when using formal financial data, no gender gap is found between male- and female-owned firms and access to credit for Sub-Saharan Africa. Hansen and Rand's (2014) research addresses some of the discrepancies in measuring gender inequality in business; but all measures are important. Self-reported measures bring light to how people view their access to finance and some of the institutional inequalities faced by women. Other characteristics including education, restrictive laws, and culture can all play a part in unequal access to finance. Formal financial data provides insight for the financial service perspective. From the financial institution there is no inequality; therefore, both measures prove significant as one could very well have an effect on the other. If there is a perceived inequality, but the gender difference doesn't occur in the formal data, then the inequality could be driven by other characteristics.

Another study conducted by Toyin Segun Ogunleye (2017) on Nigeria focuses on microfinance and the role of women in improving participation in financial services. Financial inclusion is suggested as the solution to decreasing the poverty that is especially prevalent for women living in rural areas of Nigeria. Women tend to be less risky with financial decisions and come with a lower moral hazard than men who are less cautious investors. Given this assumed behaviour, Ogunleye (2017) tests if females are more likely to repay loans given through microfinance. The findings prove that increased microfinance loans to women in Nigeria improve the repayment rate and decrease the risk involved. Increased inclusion of women in microfinance not only reduces the gender inequality created by lack of participation in financial services but also benefits lenders by increasing their repayment rate on credit loans. This relationship proves why microfinance is an efficient method of decreasing the gender gap in poverty-stricken areas.

Nwosu and Orji (2017) study the relationship between gender equality and access to formal credit in Nigeria. The authors focus specifically on the impact of formal access to credit on small- to medium-sized firm performance. The credit constraint impacts small- to medium-sized firms by reducing productivity and reinvestment into the firm. The authors find this effect to be even larger for female firms once adjusted for estimator biases. Formal access to credit is also found to have significant impact on firm performance; while informal access to credit does provide a finance option that would otherwise not exist, it is not a reliable or efficient method of increasing firm growth and productivity. Female-owned firms must be able to access formal financial services in order to grow their business and close the gender gap. The authors suggest that reforms are necessary to increase access to formal credit loans and that will in turn improve performance of SMEs, especially female-owned firms in Nigeria.

Johnson and Nino-Zarazua (2011) investigate financial access and exclusion in Kenya and Uganda, emphasising that research must be country specific to include social institutions and other unique variables. The authors find that access to formal financial services is heavily influenced by factors such as employment, education, gender, and income. However, the geographic component of rural versus urban does not have an effect on accessing financial services. The inclusion of country-specific variables allows for a detailed analysis that reflects the current level of access in Kenya and Uganda, which in turn allows for better policy analysis and implementation. The authors highlight the importance of using country-specific studies and condemn the one size fits all policy campaigns that are usually favoured by supranational organisations. Johnson and Nino-Zarazua (2011) also investigate the level of formal financing by breaking down the dataset analysis to informal, semi-formal, and formal. The different levels of financing allow for important cultural and socioeconomic factors to be captured that would not be otherwise. Gender inequality favouring men in accessing formal financial services is supported by empirical findings; however, in semi-formal and informal sectors women actually have more access than men. Again, this proves the importance of a country-specific analysis as certain cultural and social practices are vital to providing accurate results. Gender inequality in accessing financial services is dependent on the nation and its unique characteristics. The inclusion of differing levels of formality in finance, as well as country-specific variables, is necessary for impactful results.

In general, literature addressing inequality and access to financial services rarely analyses the gender difference. This leaves a major component out of the discussion on increasing access to financial services and overall development in developing nations. Women are a key component to development and empowerment process, so closing the gender gap is important for the next stage of egalitarian development in developing economies. Without the inclusion of the gender component much of the research conducted is rendered ineffective for policy creation. The existing literature on gender inequality and access to finance is limited in two major areas: statistical analysis and country-specific analysis. Authors that do focus on the gender component either do so through analysing the disparity by way of literature review and lack of statistical analysis or focus on a region and miss out on important country characteristics. Without statistical analysis it is impossible to understand the relationship between chosen variables and their magnitude on access to financial services. Therefore, to validate and deepen the understanding of the gender equality factors that limit access to financial access, quantitative analysis is compulsory for future research. Cross-country studies provide for substantial and valid research, but lack of detail leaves out important nation-specific characteristics that influence culturally driven gender inequality. In order for policies to be effective in closing the gender gap of finance then they must be based on data specific to the nation.

The focus of this chapter is on only one nation, Rwanda, and utilises statistical analysis to better understand the relationship between gender inequality and access to financial services. The main goal is to combine proven methods and reasoning from authors Sarma et al., Arestis and Dimitriades (1997), Johnson and Nino-Zarazua and Manji (2010) to create a well-rounded regression that provides statistically significant results. As used in Johnson et al., external characteristics, culture, and other major economic variables are analysed to eliminate the possibility of omitted variable bias and provide nation-specific results. The influence of inclusion and equality on development as proposed by Sarma is the centrepiece for the construction of the models. The logic of Arestis and Dimitriades (1997) and Manji (2010) is applied to critically analyse the variable selection given availability and to critically analyse the results of the regressions. The following chapter will expand on the methodology employed and the selection of variables used for the regressions.

Methods to Study Gender Inequality, Access to Finance

There are few studies that encompass all aspects of gender inequality, access to finance, and development. The role of finance and gender equality is key component of growth and development. Traditional economic growth theories emphasise the importance of a few variables as key drivers of growth: maximisation of labour, capital, and technological innovation. As the study of economics continues, researchers find the theory does not encompass all factors that influence development. Developing nations stand to benefit the most from models that investigate other factors like finance and gender equality, as they lead to steady and sustainable development. Economic growth has long been a method of decreasing inequality, but this inequality could also hinder growth in the long run. Gender inequality, specifically, holds most poor nations back from realising their economic potential. Much of the current literature focuses solely on income inequality, which is a culmination of institutional inequalities like unequal access to education, financial services, or incomplete property rights. Development is not a one size fits all scheme and it is important to include a well-rounded dataset that is country specific.

The linkages between financial access, gender equality, and development consist of a combination of growth theories and empirical studies that test the relationships between the three. The financial system plays a key role in the development of a nation both economically and qualitatively. The direction of the relationship is up for debate and should be analysed on a case by case basis, since all economies are different; however, what is not up for debate is the fact that developed economies have well-developed financial systems. According to Arestis and Dimitriades (1997) economic growth leads to financial development in over half the countries they analysed. The authors note that this relationship is not applicable to all cases and financial development could lead to economic growth. From their study the authors suggest the usage of time series data and emphasise the importance of capturing individual country characteristics. The authors note that a majority of the research conducted on the subject of economic growth and finance utilises the ratio of deposits to GDP. Thus, it can be concluded that to study the relationship between financial development and growth, the number of accounts and the GDP are commonly accepted as appropriate measures. Economic growth is an indicator of development and measuring the effect of accounts on GDP can provide the empirical linkages necessary to comment about the role of finance and development.

Financial access and development are just one aspect of development economics; the influence of finance and gender equality also influences economic growth and sustainable development. Manji (2010) provides an analysis of previous development policies and the intended expansion of financial inclusion. According to reports from the World Bank, growth of the financial is interconnected with institutional reforms, especially those concerning property rights. Increased financial access would involve a majority of the population that would otherwise remain inactive financially. Manji (2010) points out that the methods by which nations intend to increase access could inadvertently exclude more individuals, women specifically. By limiting the requirements to access, like using land titles as collateral, the barrier to entry is raised for much of the population. The policies implemented by governments have the potential to further decrease access as well as slow growth. It is important to understand that access to finance includes not only the acquiring of loans but also the usage of debit accounts. Financial institutions provide more services than just creating debt for borrowers. In developing nations, they provide a safe place to keep wages and savings, as well as acting as a financial educator. The participation by the majority in financial services has the potential to influence growth and development, not the participation of a select few. The effect of participation on growth can vary from nation to nation making the data and research methods utilise a prominent component to the viability of results.

In their research Johnson and Nino-Zarazua stress the importance of including the effects of social institutions on access to finance and inequality. Many of the measures regarding development, financial access, and institutional inequality must therefore be nation specific to be useful. The authors find that longstanding historical factors and cultural precedence contribute to gender inequality as well as financial exclusion. The linkages between financial inclusion, growth, and equality are general in logic, but in the real world these factors influence a nation's development path. As mentioned previously development policy is not one size fits all and what influences development and gender inequality varies from nation to nation. Without testing on a case by case basis, researchers run the risk of making a generalisation on the subject of development that could potentially cause harm.

Tom Goodfellow (2017) notes that Rwanda sets itself apart on the development spectrum of African nations. The nation succeeded in increasing gender equality ratings and cultivated a productive economy over the past two decades. Goodfellow (2017) suggests that for growth to continue Rwanda must utilise all available resources. Development occurs in stages with political stability, quality infrastructure, gender equality, education, and other components increasing over time. Goodfellow (2017) acknowledges Rwanda's attempt in attracting business and generating revenue through land leasing but determines that this alone is not enough and the nation must generate growth in other areas that are more sustainable. The case for an efficient and inclusive financial sector is further justified by Goodfellow's (2017) findings. Rwanda's reliance on land leasing and tax revenue does not produce enough capital necessary to overhaul the system to achieve peak productivity. An efficient financial sector allows individuals to save money or obtain loans that are then used to employ all resources and increase productivity. Right now, much of Rwanda's resources are underutilised and contribute little to their development; however, when business is supported and individuals take advantage of financial services to increase their financial security the entire economy grows. Wages are saved or invested to buy taxable assets which then produce higher tax revenues that can be reinvested into improving infrastructure, education, and attracting business both domestically and abroad. Development is not simple and numerous development authors acknowledge this fact, but they refuse to accept its complexity as a reason for nations to remain undeveloped. Every nation has a unique development path, but equality, financial access, infrastructure, and other factors increase the ability of nations to translate this to sustainable development.

In the case of Rwanda over half the population is female making equal access across genders especially important. Whether access to finance leads to growth or growth leads to access is not of major importance at the moment; what matters most is that financial inclusion indicates a qualitative level of development. It indicates that individuals are earning enough wages to participate and plan for the future, as well as having the knowledge to understand the benefits of financial access. A healthy financial sector also promotes sustainable long-term growth, as participants are able to plan for the long term financially. The gender component is of particular importance for long-run development. Nations that are inherently unequal or institutionally unequal do not develop beyond a certain point. Growth is limited when at least half of the population is unable to contribute, and gender inequality contradicts the necessary maximisation of labour and capital that create growth. Institutional equality and the opportunity for women to participate in the economy open up another level of development. Rwanda is interesting in particular as it is regarded as being fairly equal for a developing nation. About 57% of the population is female and that is reflected proportionately in their parliament. Women hold half the seats and the nation is seen as a leader in gender equality. Much of this is a direct result of the 1994 genocide, where the nation saw massive loss of life and destruction. The nation had to make women a part of rebuilding and progress towards development. This equality appears only to be surface level and does not necessarily extend to real tangible equality. Therefore, the following research seeks to understand the relationship between gender equality, access to finance, and development; and whether or not they play a significant role in the case of Rwanda.

Data

Data is the most important feature of any empirical research; therefore, equality, development, and financial inclusion must be sufficiently represented. The data on inequality and growth is sourced from the World Bank's World Development Indicators databank; and data on financial access is sourced from the International Monetary Fund's Financial Inclusion Survey databank. World Bank's world development indicator database is commonly used in the literature surrounding inequality and development. Authors like Asideu (2003), Blackden et al. (2007), and Sarma et al. source all of their institutional and equality data from the WDI databank. The International Monetary Fund financial inclusion database encompasses all Finscope and Finaccess survey data that both Aterido and Johnson and Zarazua use in their studies to determine financial access levels. The selected variables intend to encompass the following dimensions: GDP, interest rates, property rights, gender equality, and financial accounts. GDP is a common measure used in development literature and indicates growth and development levels (Arestis & Dimitriades, 1997). Since GDP incorporates private consumption, investment (both private and government), government spending, and net trade, it acts as a proxy for development levels. Developed nations have greater GDPs than developing nations, so the higher the GDP the greater the development. For these reasons it is used as the measurement of development/growth for this study. Total accounts, debit accounts, and loan accounts represent the level of financial access, and are the most available measure of financial access for the nation of Rwanda. Total accounts are used as an aggregate representation of access. Debit accounts represent non-debt incurring financial access and loan accounts represent debt incurring accounts. It is important to distinguish between the two as inequality can especially be present in the obtainment of loans. Interest rates are also of great importance especially since they reflect the monetary policies of the nation and can influence access to finance. The interest rates for both deposit accounts and loan accounts are used as they are the most relevant when investigating individual access. As mentioned earlier property rights play a role in development, equality, and access. In their most basic form property rights represent the country-specific institutional factors, but they also influence willingness to access finance. If there are laws in place to ensure the ownership rights of the people there is a greater chance that individuals will risk putting financial assets with a financial institution. Property rights are measured by the CPIA property rights and rule-based governance rating (1 = low to 6 = high) and measure the extent to which economic activity is enforced by law-based governance. Finally, to capture the element of gender equality the CPIA gender equality rating (1 = low to 6 = high) is used. This measure encompasses the institutions and programs that ensure equal access to health, education, the economy, and protection by law. This variable is preferred as it combines all components of gender equality that are scarce in data. These variables will provide a well-rounded representation of Rwanda's development, financial inclusion, and gender equality from 2004 to 2018.

Methodological Outlook to Study Gendered Finance

With the variables selected the regressions and methodology can be constructed. All regressions are held under the assumptions of the classic linear regression model and are multivariate. Ordinary Least Squares method is employed in the statistical package Eviews to measure the relationship and significance between the selected dependent variable and independent variables. Following the works of Beck et al. (2009), Manji (2010), and Johnson et al., the first regression investigates the relationship between institutional factors and financial inclusion. Sarma specifically notes that financial exclusion is the manifestation of social exclusion. Human development indicators, like gender equality and property rights, and financial inclusion are highly correlated.

$$\text{Total Accounts} = \alpha + \beta_1 \text{CPIA gender equality} + \beta_2 \text{CPIA property rights} + \mu \tag{3.1}$$

H_0: Gender equality and property rights do not influence the number of total accounts in Rwanda.
H_1: Gender equality and property rights do influence the number of total accounts in Rwanda.

In research by Beck et al. (2009) and Johnson et al. property rights and gender equality are found to greatly affect financial access, or in this case total accounts. Given the findings of these studies and others, property rights and gender equality are expected to have a positive impact on the number of total accounts. The higher the equality and property rights rating, the greater the total accounts. The variables are not squared as there is no reason to believe that they would be graphically parabolic and not linear. Unlike a variable like age, these ratings at any given level should still have positive impact on total accounts. Rwanda has an average rating of 4 for gender equality and 3.5 for property rights over a thirteen-year period. These ratings are on the upper middle end of the scale and are higher than the average for Africa, especially for Sub-Saharan Africa. The results of this regression indicate if equality and property rights do impact total accounts as expected.

The second regression aims to determine the effects of monetary policy factors on the number of total accounts. As explained earlier monetary policy is measured by the interest rate for loan and debit accounts. Suggestions by Arestis and Dimitriades (1997) for future studies include using country-specific characteristics that would influence financial services. Interest rates signal economic health and policy as well as influence consumer behaviour. The two are selected as they have the most direct impact on consumers of financial services.

$$\text{Total accounts} = \alpha + \beta_1 \text{interest rate for deposits} - \beta_2 \text{interest rate for loans} + \mu \tag{3.2}$$

H_0: Loan and deposit interest rates do not influence the number of total accounts in Rwanda.
H_1: Loan and deposit interest rates do influence the number of total accounts in Rwanda.

The relationship between interest rates and total accounts is linear in nature but the signs for the account type will differ. Deposit accounts are expected to have a positive relationship with total accounts. As interest rates increase for deposits, the incentive to open a deposit account increases. For loan accounts a negative relationship is expected. If loan interest rates increase the incentive to borrow is decreased, and loan accounts then decrease. While the dependent variable is total accounts the logic holds as well as the expected signs. There is some uncertainty when using this regression. As existing accounts are still considered the influence of the interest rates may not be as influential as assumed. The interest rate would only impact the opening of new accounts. Since the number of accounts is measured using yearly data the intuition should hold as a change in total accounts over time signals the opening of new accounts. Both interest rates are expected to influence total accounts for the nation of Rwanda.

The third regression attempts to measure the effects of numerous inequality measures on gender equality in Rwanda. Manji (2010) notes that gender inequality can result from inequality in accessing financing, therefore, both deposit and loan accounts are included to determine their significance in influencing gender equality. Johnson and Nino-Zarazua (2011), as well as many other authors explained, emphasise the importance of institutional factors like education and unemployment. Education is divided by levels: primary the lowest and tertiary the highest. Arestis and Dimitriades (1997) suggest the use of country-specific economic health variables like unemployment and GDP. The two variables signal overall economic health and how these factors influence inequality.

$$\begin{aligned}\text{Gender equality} = \alpha + \beta_1 \text{loan accounts} + \beta_2 \text{deposit accounts} \\ + \beta_3 \text{female primary enrolment} \\ + \beta_4 \text{female secondary enrolment} \\ + \beta_5 \text{female tertiary enrolment} \\ + \beta_6 \text{female unemployment} + \beta_7 \text{GDP} + \mu \end{aligned} \tag{3.3}$$

H_0: Loan and deposit accounts, education, unemployment, and GDP do not influence gender equality in Rwanda.
H_1: Loan and deposit accounts, education, unemployment, and GDP do influence gender equality in Rwanda.

Unlike regression one which measures the effect of institutional factors on the number of total accounts, regression three tests the influence of the type of accounts, female enrolment in education, female unemployment, and Rwanda's GDP. This model makes it possible to investigate the relationship between all factors of interest that could influence gender equality; however, this model is not all encompassing. While it will determine the significance of these factors in explaining gender equality in Rwanda, there are some specific components that might not be captured by the model. Given that the measure for gender equality is a rating based on institutional equality factors listed previously, the relationship between this and the explanatory variables is also indicative of development levels. Equality ratings, number of accounts, education, employment, and GDP are expected to be higher the greater the development. Regression three provides the magnitude of influence the independent variables have on gender equality, and from this gender equality as it exists in Rwanda is revealed.

The fourth and final regression attempts to measure the influence of gender equality, loan accounts, deposit accounts, interest rates on loan and deposit accounts, and property rights on GDP for Rwanda. Sarma finds that GDP predicts development levels of the nations studied and nations with low GDP also have lower literacy rates, income, and are less financially inclusive. Arestis and Dimitriades (1997) note that the relationship between the financial sector and development does not behave in a definitive direction. Therefore, measures of institutional policies and financial inclusion and their contribution to development are tested. The main motivation behind including this regression is to know the impact of these variables on GDP as it allows for linkages to be made across institutional factors, financial inclusion, and development.

$$\begin{aligned} GDP = \alpha + \beta_1 \text{gender equality} + \beta_2 \text{loan account} \\ +\beta_3 \text{deposit account} \\ +\beta_4 \text{interest rate for deposit accounts} \\ \beta_5 \text{interest rate for loan accounts} \\ +\beta_6 \text{property rights} + \mu \end{aligned} \quad (3.4)$$

H_0: Gender equality, loan accounts, deposit accounts, loan and deposit account interest rates, and property rights do not influence GDP for Rwanda.

H_1: Gender equality, loan accounts, deposit accounts, loan and deposit account interest rates, and property rights do influence GDP for Rwanda.

All of the explanatory variables in this regression are assumed to be independent of one another. It is expected that a majority of the variables influence GDP in some way. Institutional factors like gender equality and property rights are expected to have a positive impact on GDP. Financial inclusion measured by the number of loan and deposit accounts should also have a positive relationship with GDP. Loan account interest rates are expected to have a negative relationship with GDP and interest rates for deposit accounts are expected to have a positive impact on GDP.

Conclusion

Each regression is designed to determine the relationships between factors that have recently been identified as important to development. Regression one investigates the relationship between institutional factors and financial inclusion that are unique to Rwanda. Regression two tests the impact of interest rates on the number of total accounts for the nation. Regression three determines if loan and deposit accounts influence the gender equality rating. Finally, regression four seeks to understand what institutional, financial inclusion, and policy factors affect GDP. All regressions are time series multivariate linear regressions and are tested in the statistical package Eviews using the ordinary least squares method. Each one is tested for the presence of heteroskedasticity (white test) and serial correlation (lm test). The results of each regression are analysed in detail and where they fit in the greater literature is explained in the following section.

Financial Inclusion, Gender Equality, and Development in Rwanda

Development economics is multifaceted, and the development of nations is even more influenced by a variety of factors. The regressions laid out in the previous chapter are in no way exhaustive in testing every component that influences development, but they attempt to answer the question: How do gender equality, financial inclusion, monetary policy, and development interact? The purpose of having four separate regressions is so that specific relationships of interest can be better understood for Rwanda. Each component can be linked together and its relevance to development known once the results for each are obtained.

Regression 1: The Effect of Institutional Policies on Financial Inclusion

As described in this chapter the purpose of regression one is to determine if institutional factors, measured by gender equality and property rights, affect total accounts in Rwanda.

As can be seen in Table 3.1 both gender equality and property rights are statistically significant in relation to total accounts at 5%. The control c is the base level of accounts if all other variables were set to zero; meaning that if gender equality and property rights were set to zero there would be a negative relationship between the constant and total accounts. For every one-unit increase in the control there is a 1,696,622 decrease in total accounts. The standard error of the control is large however taking the t-statistic and probability into account it is possible to conclude that the constant is statistically significant at all levels. Now moving on to the variables of interest, gender equality's relationship with total accounts is positive as expected. For every one-unit increase in gender equality there is a 2,186,687 increase in total accounts and this relationship is statistically significant at a 95% confidence interval. Gender equality does in fact influence the number of accounts in the nation of Rwanda. It is important to note that this is not completely indicative of equality in access to finance on the basis of gender, but instead signals that gender equality in Rwanda does influence the number of accounts held by the population. However, the CPIA gender equality index does take institutional factors like equal access to education, the economy, healthcare, and protection under the law. It is therefore possible to infer that this interaction does suggest

Table 3.1 Results of effects of institutional factors on financial inclusion

Dep var. total accounts	
Control	-16966622***
	(675852)
Gender equality	2186687**
	(872766.6)
Property rights	3537168**
	(1344012)
R^2	0.879
Adjusted R^2	0.852
Observations	12
Standard errors in parentheses	
*** significant at 1%	
** significant at 5%	
* significant at 10%	

relatively equal access by gender to opening an account. Authors Batuo et al. (2010), Jayachandran (2015), and Johnson and Zarazua all find that government policies that increase gender equality experience a significant increase of financial inclusion in African nations. Discriminatory social institutions reduce access to finance and the findings for Rwanda support the results of existing literature.

The next variable of interest is property rights; which indicates the extent to which private economic activity is facilitated by an effective legal system. As can be observed from Table 3.1 property rights are statistically significant at a 95% confidence interval; and for every one-unit increase in property rights there is a 3,537,168 increase in total accounts. Therefore, the null hypothesis for regression one, property rights and gender equality do not influence total accounts, can be rejected. The results fully support findings by Beck et al. (2009), Johnson and Zarazua, and Aterido et al. who note that a strong legal system with specific laws regarding property rights increases financial participation and reduces inequality. Smooth development over time occurs when property, equality, and the right to actively engage with financial services are guaranteed by law. Property rights play an integral role in consumer attitudes and risk-taking behaviour. Financial risk and trust of financial institutions are motivated by the security individuals are guaranteed. Property rights given by law protect the people from any malicious behaviour on the side of financial institutions. Even without an overt contract regarding the transaction between financial institutions and individuals, property rights ensure that there is a base contract that must be upheld under general law. Laws specific to the financial industry provide further protection for the consumer and are a key indicator of development.

There are a few key components that ensure the soundness of the model; the first is adjusted R^2. Adjusted R^2 is preferred over R^2 as it adjusts as more explanatory variables are added to the model. In regression one the difference between adjusted R^2 and R^2 is small with the values of each being 0.85 and 0.87 respectively. An R^2 close to one is desired as it proves the explanatory power of the independent variables and the strength of the model. The sum squared residuals are large at 4.31E+12 and suggest that there are other random variables besides gender equality and property rights that could influence the number of accounts in Rwanda. This is highly likely as a number of factors influence not only the availability of access but the consumer behaviour that determines access. While it would be preferred to incorporate all variables that influence total accounts, it is near impossible given limited availability of data and the fact that some factors are unable to be measured. The next component is the F-statistic and the probability of the F-statistic which captures whether the independent variables, property rights, and gender equality are significant in explaining total accounts. The F value is 32.6, and, while low, is statistically significant at all levels. The lower F value suggests that property rights and gender equality do explain a portion of total accounts, however, they do not capture all of the factors that influence total accounts in Rwanda. The final piece of information that determines if the model is statistically sound is the Durbin-Watson statistic. The desired value for the Durbin-Watson is 2, however for this regression it is 1.26 suggesting that there may be evidence of some serial correlation.

Given a Durbin-Watson statistic of 1.26 a serial correlation LM test is performed to determine if there is in fact serial correlation. Once the test is performed with the standard 2 lags it is possible to reject the presence of serial correlation at all levels of significance. While the Durbin-Watson is not ideal the model itself is clear of any serial correlation. Next to further ensure the structural integrity of the model a White test is completed to determine the presence of heteroskedasticity. After running the White test, it is possible to reject any presence of heteroskedasticity at all significance levels.

When governments have policies in place to ensure the fair and equal treatment of consumers when accessing financial institutions, overall trust of financial services increases and in turn increases access and development. The immediate effects of gender equality and property rights are blatantly obvious; however, their effects are far reaching and influence even the smallest areas of the economy like consumer behaviour. When a nation promotes equality and protects it citizens under the law, it proves the attainment of a certain level of development and breeds opportunity for future growth.

Regression 2: Influence of Interest Rates on Financial Inclusion

The next regression models the relationship between deposit and loan account interest rates and the number of total accounts. Interest rates reflect Rwanda's monetary policy and changes in interest rates influence consumer behaviour.

The results of the OLS regression can be observed in Table 3.2. The interest rates on deposit accounts are statistically significant at 10% only and have a negative relationship with total accounts. Given the coefficient it is possible to state that a one-unit increase would result in a 4,454,923.5 decrease in total accounts. The negative relationship is expected as a higher interest rate would be beneficial for account holders and should therefore increase the number of accounts overall. Traditionally debit accounts do not accrue interest; however, this measure also includes savings accounts. Since each account has a specific set of conditions it is challenging to

Table 3.2 Results of interest rates and financial inclusion

Dep var. total accounts	
Control	-48884994***
	(8317554)
Deposit interest rates	-454923.5*
	(215227)
Loan interest rates	3330695***
	(476998.8)
R^2	0.827
Adjusted R^2	0.798
Observations	15
Standard errors in parentheses	
*** significant at 1%	
** significant at 5%	
* significant at 10%	

generalise and make an assumption on the entirety of debit accounts in Rwanda and their interest rates for analysis. However, the data is at the national level and therefore all individual characteristics are held constant. The negative relationship could be a result of some unexplained factors in the model or factors not captured in the model. The finding of importance is that debit interest rates are significant and do influence total accounts.

Loan interest rates are statistically significant at all confidence levels and the relationship between total accounts and loan rates suggests that a one-unit increase in rates results in a 3,330,695 increase in total accounts. A negative relationship is unexpected as one assumes that as interest rates increase for loans that individuals would be less likely to open new accounts at high rates. There are other factors that influence loan rates with credit-worthiness and collateral of the borrowers being large determinants. The model, however, uses national statistics and the individual characteristics should not necessarily influence the relationship of rates and accounts. Before drawing definitive conclusions on the implications of the model, the characteristics of the overall regression must be taken into account.

The adjusted R^2 is close to one at 0.797 and suggests that the model's independent variables do have power in predicting total accounts. The sum squared residuals are large at 1.23E+13 as can be expected given the nature of the variables and simplicity of the model. Much of the residuals lay outside of the regression line and exist outside of the model. Since it is difficult to capture all factors that influence interest rate and total accounts with limited data availability, a higher sum squared residuals is not a cause for alarm. The F-statistic is low at 28.59 given the raw data values and nature of the data but is statistically significant at all levels. Interest rates for loan and debit accounts are not overwhelmingly significant in explaining total accounts, but they are a component of total accounts. Finally, the Durbin and Watson statistic takes a value of 1.20 when rounded, suggesting the possibility of positive serial correlation in the residuals. The results of the serial correlation LM test determine that there is no serial correlation in the model. The White test also confirms an absence of heteroskedasticity and equal variability in the model's observations.

The regression output proves two valuable pieces of information. First, the independent variables are statistically significant and influence the number of total accounts in Rwanda. The unique findings support Johnson and Zarazua's proposition of country-specific data on financial markets. Other authors like Arestis and Dimitriades (1997) and Ang (2010) find that well-developed and efficient financial services greatly increase inclusion and decrease inequality for developing nations. Interest rates are indicators of the quality and efficiency of financial services and their accessibility to the people of a developing nation. The results of regression two highlight the importance of interest rates and their impact on access. However, the explanatory power of loan and deposit interest rates is limited and does not fully capture what affects total accounts. Model two limits the number of included explanatory variables to observe isolated effects of interest rates on total accounts to observed. The relationships are unexpected but much of this could be due to the limited nature of the model. It is impossible to suggest that the model includes all aspects of Rwanda's monetary policy, but it provides a snapshot of a very small aspect that affects individuals and their motivation to access finance. Overall, the significance of the results is minimal in the whole of financial inclusion and development but reveals some important information on access in Rwanda.

Regression 3: Effect of Economic and Institutional Factors on Gender Equality

Gender equality and access to finance itself are not a deeply researched topic, yet the implications of unequal access can severely impact a nation's development path. The ability to model and measure gender equality and financial inclusion in detail is unfortunately not possible given the available data. It is instead included as a feature in the larger analysis of financial inclusion and development for Rwanda. Regression three differs from regression one by determining if the number of accounts (access) and other proxies used for gender equality influence the gender equality rating of Rwanda. Variables typically used to indicate specific areas of inequality are limited; however, data on education and unemployment is available and is included to encompass possible influences on gender equality. The effect of access along with other equality proxies is measured on gender equality to understand if the instance of increased financial inclusion improves gender equality overall. Accounts are broken up by loan and deposit to get an idea of how different account types influence gender equality and the same is done for education. Depending on the results of the regression it is possible to hypothesise the circumstances surrounding the impact of decisive influences on equality and financial inclusion on equality ratings (Table 3.3).

The results are telling with all but one variable being statistically significant at a 90% confidence interval. The null hypothesis stated in this

Table 3.3 Results of economic and institutional factors on gender equality

Dep var. gender equality	
Control	-23.49**
	(1.613)
Loan accounts	-7.49E-06**
	(5.19E-07)
Deposit accounts	-1.10E-06**
	(7.10E-08)
Primary	0.113**
	(0.009)
Secondary	0.014
	(0.013)
Tertiary	0.565*
	(0.075)
Unemployment	2.476*
	(0.267)
GDP	1.55E-09**
	(7.12E-09)
R^2	0.999
Adjusted R^2	0.997
Observations	9
Standard errors in parentheses	
*** significant at 1%	
** significant at 5%	
* significant at 10%	

chapter can therefore be rejected for all variables but secondary education. Loan accounts and deposit accounts are significant at 5% but both are negatively related to gender equality. These findings completely contradict all results posed by Sarma et al., van Staveren (2001), Batuo et al. (2010), and Ang (2010). All research conducted on the relationship between financial access and gender equality comes to the conclusion that there is a positive relationship between the two. Rwanda's characteristics differ from the previous nations studied and the higher instance of equality could contribute to the unexpected relationship. It is possible that the gender equality rating decreases as total accounts increase; however, it is difficult to explain why this occurs. While the relationship is negative the impact of the relationship is quite small with a one-unit increase in loan accounts resulting in a 0.00000749 decrease in gender equality. A one-unit increase in deposit accounts results in 0.0000011 decrease in gender equality. Since the results are not supported by any previous findings and the relationship contradicts nearly all empirical literature, the impact of the results is inconclusive in regard to the wider literature.

Primary education is significant at a 95% confidence interval and a 1% increase in primary education for females results in a 0.113 unit increase in gender equality. Secondary education is not statistically significant at any accepted level therefore failing to reject the null. Tertiary education is significant at a 90% confidence interval and for every 1% increase in female tertiary education gender equality increases by 0.565 units. Primary education and tertiary education are the two major influencers of gender equality; however, secondary education fails to impact the dependent variable. A majority of females in Rwanda are educated at the primary level but enrolment drops off at the secondary level and continues a downward trend for tertiary enrolment. The insignificance of secondary school could suggest a gap in impact for schooling levels. The difference between an individual without a primary education and an individual with a primary education is potentially greater than the difference between an individual who has a secondary education and one who only has a primary education. The value added by secondary school may not be enough to influence gender equality for Rwanda. Tertiary education does impact gender equality and the same intuition relates to the difference between tertiary and secondary. Since tertiary is the highest level of education, female enrolment would be the most impactful on equality. Women and girls tend to receive less higher level education so the magnitude of an increase in tertiary enrolment is the most effective in increasing gender equality. van Staveren (2001), Jayachandran (2015), Aterido et al. (2013), and Johnson and Zarazua all find that external characteristics like education and unemployment create inequality between genders and these inequalities show up as unequal access to finance. According to the aforementioned authors education, employment, and economic participation heavily impact the levels of gender equality for a nation and in turn influence access and development. The results suggest that the base level of education and the highest level of education are the most significant in influencing gender equality.

The female unemployment rate is significant at 10% but suggests a positive relationship with gender equality. Authors Johnson and Nino-Zarazua (2011) found that factors like employment, education, gender, and income influence the inequality observed in access to finance. The same logic is applied in the inclusion of the independent variables; however, the relationship with unemployment and gender equality is the opposite of the majority of literature on gender equality and development. Unemployment should have a negative relationship with gender equality as the ability and participation of women in the workforce are believed to be a feature of equality between the genders. Just because the relationship is not what is expected does not mean that the result does not indicate other occurrences. A positive relationship between unemployment and gender equality could suggest a perpetual occurrence of poverty for Rwanda. The behaviour is common in development research during instances of longstanding poverty. If poverty and unemployment have existed for a long period of time, then eventually other factors begin to evolve regardless of a key economic development indicator. In the case of Rwanda, society has become more equal over time but unemployment across both genders has remained relatively equal with male unemployment rates occasionally greater than female. Rwanda could very well be a special case were perpetual poverty mixed with equal unemployment values for both genders render unemployment a positive effect on gender equality.

GDP as a measure of development is statistically significant at a 95% confidence interval but is limited in magnitude. The effect of GDP on gender equality is minimal and the main connection between the two are higher values for more developed nations. Sarma et al. find that institutional factors and development levels influence the instances of inequality in a nation. Countries with higher GDPs typically have higher gender equality ratings as they are more developed. The level of development must be considered as being dependent on gender equality and vice versa. Development does definitively cause higher instances of gender equality and gender equality does not singularly cause increases in development; instead the effects flow from both directions. Even though the relationship is small it does prove the point that a higher GDP is associated with gender equality. Economic performance is linked directly to the equal inclusion of half the population. The third model proves that country-specific characteristics like education, unemployment, accounts, and GDP allow for relationships to be overserved that might not otherwise be if studied on a larger scale.

The adjusted R^2 is almost perfect at 0.997 and the sum squared residuals are minimal at 0.00036. Therefore, the model has strong predictive power with minimal random residuals. The F-statistic is large at 352.91 and is statistically significant at 5%. While it is preferred that the F-statistic is significant at all levels, this still proves the significance of the independent variables in explaining gender equality. The Durbin-Watson statistic is high at 3.12 but after running the serial correlation LM test, no evidence of serial correlation is found. Even with serial correlation ruled out the high Durbin-Watson makes sense given the nature of the variables and it is expected that the values from the previous time period would influence the value of the variables in the next period. A White test is performed to ensure the quality of the model and after conducting a White test no evidence of heteroskedasticity is found. Overall regression three proves some important points on gender equality in Rwanda. The first is the very small, almost negligible, negative influence of loan and deposit accounts on gender equality. Next is that education, especially tertiary, significantly impacts gender equality. Unemployment is also significant but does not behave in a normal way, suggesting that longstanding poverty and relatively equal unemployment between genders are unique country characteristics that change the behaviour in regard to gender equality. The last point proves that higher GDP, and therefore development, influences gender equality in a positive way. Just with these findings alone, the importance of individual country characteristics is realised. Without analysing the factors at the national level important observations and relationship could not be captured and instead be lost.

Regression 4: Influence of Financial Inclusion and Institutional Factors on Development

The fourth and final regression encompasses all previous variables and regresses them against GDP so that the total effects on development can be observed. Rwanda's development is the result of numerous factors that could not possibly be fully captured by this model. The goal of the fourth model is to understand how monetary, institutional, and access factors influence growth and development as a whole. Since the model captures only a few components of changes in GDP over time, some of the results may not be significant; however, by modelling certain variables in smaller regressions the individual effects are still addressed.

Table 3.4 depicts the estimation output for the final regression. Only three variables are statistically significant at any accepted level. Gender equality is significant at a 95% confidence level and has a positive relationship with GDP. These findings support similar results in Sarma et al. that determine the significant impact of social factors on development. The authors find that instances of unequal societies largely reduce the GDP of

Table 3.4 Results of financial inclusion and institutional factors on development

Dep var. GDP	
Control	2.86E+10
	(1.69E+10)
Gender equality	1.72E+09**
	(6.32E+08)
Loan accounts	8570.119**
	(2866.056)
Deposit accounts	660.931*
	(322.552)
Loan interest rate	-1.47E+09
	(8.70E+08)
Deposit interest rate	1.14E+08
	(1.17E+08)
Property rights	-2.73E+09
	(1.84E+09)
R^2	0.983
Adjusted R^2	0.962
Observations	12

Standard errors in parentheses
*** significant at 1%
** significant at 5%
* significant at 10%

developing nations. The magnitude of gender equality's effect on Rwanda's GDP is very large. As mentioned in regression 3, well-developed nations have higher instances of gender equality. Rwanda's GDP significantly benefits from an already high-level gender equality but could benefit even more by increasing factors that go into measuring the gender equality index variable. Equality is a major component of a country's development path and without an equal society growth is stunted.

Loan accounts are statistically significant at 5% and positively impact GDP. A one-unit increase in loan accounts increases GDP by 8570.12. Deposit accounts are also statistically significant at 10% and a one-unit increase would result in a 660.93 increase in GDP. The magnitude of both variables reveals the importance of access to finance. Any increase to accounts has a major influence on the GDP and therefore development of Rwanda. As more people are financially included the overall economy will begin to pick up and GDP will increase as a result. As Arestis and Dimitriades (1997) explained, the financial system plays a key role in the development of a nation's economy. The two are mutually beneficial, creating a recurring increase of GDP and accounts as the other increases.

Interest rates on deposit and loan accounts are insignificant at all levels and do not influence GDP, thus, supporting the findings of Sarma et al. Regression two suggests that interest rates influence the number of total accounts however those findings do not translate to this model. Even if the variables were significant their impact would be minimal and would not be considered a powerful explanatory variable. Property rights are also insignificant at all levels and the magnitude of impact on GDP would be minimal as well. It is logical to assume more property rights would lead to a more developed nation with a higher GDP, but for Rwanda this is not necessarily true. Property rights have remained consistent over the time period and without any major changes in the data, the consistency renders it insignificant. If there are no changes over time in property rights and GDP continues to grow then a lack of significance is not surprising. These variables have no explanatory power, but gender equality and accounts significantly explain a good portion of GDP.

The model is strong and captures the effects efficiently. The adjusted R^2 is right at 0.96 and the F-statistic is large and significant at all levels. The Durbin-Watson is perfect at 2 and after running the serial correlation LM test any chance of serial correlation is rejected. The White test is conducted to determine homoskedasticity of the model and this is confirmed with zero evidence of heteroskedasticity. The model is well fitted, and gender equality and accounts significantly explain changes in GDP. The fourth model ties in all components and reinforces previous findings. Gender equality and access to finance heavily influence development. Since the genocide Rwanda has outpaced comparable LDCs in both gender equality and number of accounts. Developing nations must make equality and financial inclusion a priority if they want to break through the development ceiling.

Results

The four linear regressions set out to determine the interactivity of gender equality, financial inclusion, and development. Development economics has evolved to highlight gender equality and financial inclusion as two major components of long-run sustainable development. Regression one sufficiently proves the significant and positive relationship institutional equalities have with the number of total accounts in Rwanda. Protection of property and equality in education, healthcare, law, and the economy not only provides the opportunity for more accounts to be opened but protects the rights of those opening the accounts. Regression two evaluates the effect of interest rates on loan and deposit accounts on the number of total accounts. Both interest rates are significant; however, their signs are opposite the expected relationships. Individual country characteristics not captured by regression two could influence the signs observed, but overall the model proves the importance of interest rates in accessing finance. Regression three captures the factors that most influence gender equality ratings in Rwanda. Education, accounts, unemployment, and GDP all significantly influence gender equality, reinforcing the idea that better access to education, finance, and a higher GDP makes society more equal. Unemployment positively impacts gender equality, but longstanding poverty levels and equal unemployment between men and women could be abnormalities that make this relationship unique to Rwanda. Regression four ties together all three regressions and proves the significant role gender equality and access to finance play in development. All four regressions support numerous findings in the current literature surrounding development, equality, and finance; thus, validating the results and their contribution to the understanding of development in Rwanda.

Conclusion

The findings of the regressions are consistent with much of the literature regarding inclusion, equality, and development. The findings from regression one prove gender-based barriers impede economic growth, where poverty is the highest among developing nations. Any sort of gender-based obstacle like exclusion from employment, education, or the economy has a detrimental impact on developing nations. The role of interest rates in Rwanda's access to finance is determined as significant in regression two; however, the relationship proves unique with unexpected relationships. The results from regression three and four prove that institutional factors do heavily influence equality and equality in turn influences growth and development. The requirement of equality for development does not apply exclusively to the individual and household levels; equal opportunity and access to finance must be present in firm funding. While there is no data regarding the firm-level access to finance for Rwanda, the assumption that equality and development are positively related holds. If firms are given financing based on their innovation and potential to perform regardless of gender then the economy will benefit and grow to a higher stage of development. Finally, the analysis of gender equality, financial inclusion, and development must be conducted on a case by case basis as suggested by Johnson et al. Some of the regression results are surprising but can be understood by applying country characteristics to the analysis. Overall, the results support the findings of major economic research in the fields of development and equality and contribute to the literature from Rwanda's perspective.

Methods of development have constantly evolved over time and the factors that most affect development are finally coming to light. Development is a complex process that is unique to each individual country. The components that influence development range depending on the nation and there is no exact method that could spur growth for every country. Researchers have made an effort to determine some of the most important factors of development so that analysis for individual nations proves more lucrative. Two main factors that seem to affect every developing and developed nation in the world are financial inclusion and gender equality. Both variables are influenced by the unique social and cultural aspects of each country, but their impact remains a standard for all. The goal of this research was to understand the relationships between gender equality, financial inclusion, and development of Rwanda. The influence of financial inclusion and equality on Rwanda's development proves to be significant and positive. The nation is already ahead of its counterparts in gender equality and continues to enhance their financial inclusion; but the impact of the factors is central to their development thus far.

While, the relationship proves to be significant there are some limitations to this research which lead to opportunities for improvements of future studies. Financial data for Rwanda is limited and lacks the detail necessary for deeper analysis of equality and access. Given the collection of more data, both financial and equality, it would be possible to draw stronger conclusions for the purpose of policy construction. Nations stuck in the development trap often fail to recognise the factors that would allow them to escape. The continuation of research in gender equality, financial access, and development is necessary for lesser developed nations to catch up. The world is changing, and the understanding and analysis of human decisions are more important now than ever before. Development must be tailored to the specific challenges of each nation, but financial access and gender equality will remain as major influencers of development.

Appendix

Variable statistics

Variable	*Description*	*Obs.*	*Mean*	*Std. dev.*	*Max*	*Min*
GDP	GDP in constant 2011 USD (billions)	14	6.46	2.08	10.20	3.55
Gender equality	CPIA gender equality index (1=low, 6=high)	12	3.92	0.40	4.5	3.5
Property rights	CPIA property rights index (1=low, 6=high)	12	3.31	0.25	3.5	3
Primary	Female primary enrolment (% gross)	13	142.79	5.71	149.92	132.76
Secondary	Female secondary enrolment (% gross)	13	30.44	10.79	42.23	14.13
Tertiary	Female tertiary enrolment (% gross)	10	5.80	1.63	7.18	1.99
Unemployment	Female unemployment (%, ILO estimate)	14	0.96	0.21	1.25	0.61
Total accounts	Total number of deposit and Laon accounts (thousands)	14	3145.97	2250.62	7111.02	40.40
Loan accounts	Total loan accounts (thousands)	14	215.13	170.71	484.36	2.66
Deposit accounts	Total deposit accounts (thousands)	14	2930.85	2085.83	6626.65	37.75
Loan interest rates	Loan interest rate (%)	14	16.69	0.57	17.33	15.78
Deposit interest rates	Deposit interest rate (%)	14	7.80	1.26	9.92	5.39

References

Ang, J. B. (2010). Finance and Inequality: The Case of India. *Southern Economic Journal, 76*(3), 738–761. https://doi.org/10.4284/sej.2010.76.3.738

Arestis, P., & Dimitriades, P. (1997). Financial Development and Economic Growth: Assessing the Evidence*. *The Economic Journal, 107*(442), 783–799. https://doi.org/10.1111/j.1468-0297.1997.tb00043.x

Asiedu, E., Kalonda-Kanyama, I., Ndikumana, L., & Nti-Addae, A. (2003). Access to Credit by Firms in Sub-Saharan Africa: How Relevant Is Gender? *American Economic Review, 103*(3), 293–297. https://doi.org/10.1257/aer.103.3.293

Aterido, R., Beck, T., & Iacovone, L. (2013). Access to Finance in Sub-Saharan Africa: Is There a Gender Gap? *World Development, 47*, 102–120. https://doi.org/10.1016/j.worlddev.2013.02.013

Batuo, M. E., Guidi, F., & Mlambo, K. (2010). Financial Development and Income Inequality: Evidence from African Countries, n.d., 28.

Beck, T., Demirguc-Kunt, A., & Honohan, P. (2009). Access to Financial Services: Measurement, Impact, and Policies. *The World Bank Research Observer, 24*(1), 119–145. https://doi.org/10.1093/wbro/lkn008

Blackden, M., Canagarajah, S., Klasen, S., & Lawson, D. (2007). Gender and Growth in Sub-Saharan Africa: Issues and Evidence. In G. Mavrotas & A. Shorrocks (Eds.), *Advancing Development* (pp. 349–370). Palgrave Macmillan UK. https://doi.org/10.1057/9780230801462_19

Chavan, P. (2008). Gender Inequality in Banking Services. *Economic and Political Weekly, 43*(47), 18–21.

Goodfellow, T. (2017). Taxing Property in a Neo-Developmental State: The Politics of Urban Land Value Capture in Rwanda and Ethiopia. *African Affairs, 116*, 549–572. https://doi.org/10.1093/afraf/adx020

Hansen, H., & Rand, J. (2014). Estimates of Gender Differences in Firm's Access to Credit in Sub-Saharan Africa. *Economics Letters, 123*(3), 374–377. https://doi.org/10.1016/j.econlet.2014.04.001

Issac, J. (2014). *'Expanding Women's Access to Financial Services'. 201*. World Bank. Accessed June 7, 2019, from http://projects-beta.worldbank.org/en/results/2013/04/01/banking-on-women-extending-womens-access-to-financial-services.

Jayachandran, S. (2015). The Roots of Gender Inequality in Developing Countries. *Annual Review of Economics, 7*(1), 63–88. https://doi.org/10.1146/annurev-economics-080614-115404

Johnson, S., & Nino-Zarazua, M. (2011). Financial Access and Exclusion in Kenya and Uganda. *Journal of Development Studies, 47*(3), 475–496. https://doi.org/10.1080/00220388.2010.492857

Kabeer, N. (2005). Is Microfinance a "Magic Bullet" for Women's Empowerment? Analysis of Findings from South Asia. *Economic and Political Weekly, 40*(44), 4709–4718.

Klapper, L. F., & Parker, S. C. (2011). Gender and the Business Environment for New Firm Creation. *The World Bank Research Observer, 26*(2), 237–257. https://doi.org/10.1093/wbro/lkp032

Manji, A. (2010). Eliminating Poverty? "Financial Inclusion", Access to Land, and Gender Equality in International Development: Eliminating Poverty? *The Modern Law Review, 73*(6), 985–1004. https://doi.org/10.1111/j.1468-2230.2010.00827.x

Nwosu, E. O., & Orji, A. (2017). Addressing Poverty and Gender Inequality through Access to Formal Credit and Enhanced Enterprise Performance in Nigeria: An Empirical Investigation: Addressing Poverty and Gender Inequality. *African Development Review, 29*(S1), 56–72. https://doi.org/10.1111/1467-8268.12233

Ogunleye, T. S. (2017). Financial Inclusion and the Role of Women in Nigeria: Financial Inclusion. *African Development Review, 29*(2), 249–258. https://doi.org/10.1111/1467-8268.12254

Park, C.-Y., & Mercado, R. J. (2015). Financial Inclusion, Poverty, and Income Inequality in Developing Asia. *SSRN Electronic Journal.* https://doi.org/10.2139/ssrn.2558936

van Staveren, I. (2001). Gender Biases in Finance. *Gender and Development, 9*(1), 9–17. https://doi.org/10.1080/13552070127734

CHAPTER 4

Financial Inclusion, Gender Gaps, and Development in Rwanda

Madelin O'Toole and Bhabani Shankar Nayak

Introduction

The ideas of financial inclusion and gender equality are central to economic growth and development. Both the WEF Gender Equality Report (2020) and Demirgüç-Kunt (2017) highlight financial inclusion as a method to reduce inequality and create long-term development in developing nations. Much of the current research, including Pallavi Chavan (2008), John Issac (2014), Seema Jayachandran (2015), Van Staveren (2001), Blackden et al. (2007), and others, suggests that already present inequalities in education, healthcare, labour, and wages create inequality in financial access. Asiedu et al. (2013), Hansen and Rand (2014), Nwosu and Orji (2017), and Klapper and Parker (2011) find similar results when investigating female-owned or managed firms. From the current empirical

M. O'Toole (✉)
Federal Research Division, Library of Congress, Washington, DC, USA

B. S. Nayak
University for the Creative Arts, Epsom, UK
e-mail: bhabani.nayak@uca.ac.uk

B. S. Nayak (ed.), *Political Economy of Gender and Development in Africa*, https://doi.org/10.1007/978-3-031-18829-9_4

findings, socio-cultural practices that create inequality are what cause unequal financial access.

Much of the literature surrounding female financial inclusion and development analyses multiple countries; however, Arestis and Demetriades (1997) and Johnson and Nino-Zarazua (2011) emphasise the importance of single-country analysis over time. Using time-series OLS regression, the relationship between equality, inclusion, and development in Rwanda is examined to capture unique country characteristics. Three regressions investigate the following relationships: gender equality dimensions and female account ownership; female account ownership type and economic development; and female account ownership, gender equality, and economic development.

The results of this chapter seek to set a precedent for single-country analysis that captures the unique characteristics of a developing nation. Each country requires policies that optimise the nation's economy, people, and resources. The findings of this analysis suggest that female account ownership has a significant and positive impact on economic development in Rwanda. However, gender equality measures do not have the expected relationship with female account ownership or economic development. Thus, this insignificant relationship between female financial access, economic development, and areas like education, literacy, and female parliamentary representation proves that country-specific analysis is critical to creating successful development policies.

Financial Access and Development

Financial access changes consumer behaviour surrounding personal finance. A well-functioning financial system that is inclusive of all income levels and genders promotes long-term financial thinking. Financial institutions allow consumers to store money securely, save money for the future, and send and receive payment, all of which create the economic stability necessary for long-term steady development.

Demirgüç-Kunt (2017) note the benefits of financial inclusion for economic development and its role in reducing poverty and economic uncertainty. The Global Findex Report analyses the current level of financial inclusion and proposes financial access as a dimension of sustained development. The authors (2017) emphasise the positive change in financial behaviour when access increases as a critical component in spurring qualitative development. The time horizon changes from the short term to the long term increase a consumer's propensity to save and invest in education, business, farm equipment, and other quality-of-life improvements. According to Demirgüç-Kunt (2017), between 2014 and 2017, 515 million adults opened accounts at financial institutions: in developing nations only 63% of adults hold accounts, as opposed to 94% in high-income nations (Demirgüç-Kunt, 2017). The authors suggest that engrained saving behaviours in high-income nations would become normalised in low-income countries with the availability of services. As financial access increases, low-income nations would promote a steady upward path to development through savings and investment behaviours.

On the topic of access and development, Arestis and Demetriades (1997) highlight concerns about financial development and its role in the overall development of poorer nations. The link between finance and development is heavily dependent on the characteristics of the country and its institutions. Additionally, the direction of the relationship between finance and development cannot be defined. Development and finance are interdependent, making the effects of the causal relationship and magnitude unclear. The authors, therefore, stress the importance of analysing each nation on an individual basis, as opposed to analysing multiple countries, as it captures the circumstances that influence the financial system of each nation. The results of their efforts do not allow for conclusive results that apply to all nations. Instead, the authors' results guide the methodology of future studies, stressing the importance of utilising time-series data for individual countries.

Sarma et al. discover the socio-cultural and institutional factors influence access, therefore, proving development hinges on inclusion and equality. The results of the cross-sectional analysis of 49 countries suggest that financial exclusion is a result of social exclusion. Development and financial inclusion are positively correlated, and social factors play a pivotal role in both access and development. Inequality and low access determine the development path of LDCs, proving that unequal societies will always be limited in development. Inclusion in both the social and economic dimensions promotes and sustains long-term development. Sarma et al. conclude that inclusion and human development move together at the country level. Other dimensions like education, income, and infrastructure play a role in increased access as well as development.

Economic Inequality and Financial Access

In many developing countries income inequality severely slows down growth. Corruption, commodity-based markets, and other factors create a divergence between incomes where a majority of people exist in the low-income bracket. Developing economies are limited in infrastructure, capital, financial services, all of which preserve economic disparity. Limited access to finance concentrates economic opportunity to a small group of people in developing nations. Increased access allows everyone to save and plan their economic futures, in turn, creating steady economic growth. Increased access and economic opportunity between income levels and gender offer a break in cyclical poverty. The critical behaviours of saving and investing promote long-term economic thinking and reduce short-term behaviours that continue the poverty cycle.

Demirgüç-Kunt (2017) find that higher-income individuals are more likely to have access to an account than low-income individuals. Developed countries do not experience a gap in access as the number of accounts between high and low incomes is nearly equal. However, the authors note that the gap reaches into double digits with an average of 20 percentage points in developing nations (Demirgüç-Kunt, 2017). The gap size varies by country to country, but there is a definitive gap between genders and income levels. In 2017, 50% of Rwandan adults held an account. The gap between high- and low-income individuals was 19 percentage points, and the gap between male and female was 11 percentage points. The authors note that the gaps across individual economies from 2011 to 2017 remained constant (Demirgüç-Kunt, 2017). Increased financial access and account ownership grant all individuals the opportunity to determine their economic futures as opposed to a powerful few.

Beck et al. (2009) found that the lack of access drives economic inequality and hinders future growth. The authors argue that government policies promoting financial and social inclusion are vital to closing the equality gap. They suggest that infrastructure, protection of property rights, and a competitive financial sector are essential changes in policy (Beck et al., 2009). While this research does not directly address the gender component, it does provide insights into essential policy requirements for solving general inequality. The authors highlight the importance of property rights and infrastructure, both critical components for ensuring equal access and participation between genders. Institutions like education, employment opportunity, and property rights all create a level playing field between genders. Carefully constructed policies are necessary to ensure economic equality between income as well as between genders.

James Ang (2010) analyses the relationship between financial development and policy reform on the Gini coefficient in India for over 50 years. Underdeveloped financial services increase income inequality, which substantially and negatively impacts the poor. The presence of well-developed and efficient financial services contributes to increased equality, growth, and development. While this research focuses on general income inequality and access to financial services, it does provide some insight into the subject of gender inequality. Women in developing nations already face lower wages or are engaged in unpaid work, therefore impacting them more severely than the population as a whole. If income inequality decreases as access to financial services increases, then the inequality between genders could also decrease. Again, a country's policies and institutions play a role in producing positive spill-overs necessary to equal access. Ang (2010) reinforces the importance of analysing individual country characteristics and policies to achieve successful development and increased access.

Batuo et al. (2010) test the relationship between the measure of inequality, the Gini coefficient, and financial development in Africa. The authors seek to determine if the relationship is linear or an inverted U-shape, which gives insight to the behaviour of development and inequality variables. The authors determine there is a negative linear relationship between finance and Gini coefficients, meaning that as access to financial services increases, income inequality decreases. There is no U-shape detected; therefore, financial development is to continuously reduce income inequality over time (Batuo et al., 2010). The findings suggest that policy in African nations must expand access to financial services to close the inequality gap.

Gender Inequality and Financial Access

Modern development relies on the usage of financial inclusion to reduce income inequality and sustain steady growth. While equal access between income levels is a critical element, the influence of gender inequality is even more significant. In a majority of nations, 50% of the population is female; however, governments and policies do not represent this statistic. The rights of women are secondary to the economic advancement of men in both developing and developed nations. The disparity between genders increases as income decreases, meaning that women suffer from poverty more than men. The discussion of access changes slightly when distinguishing between genders. Achieving equal economic opportunity and access between genders poses the most monumental challenge to developing nations. However, the benefits of equality not only support the economic success of women but the success of the entire nation.

Inequality and Access in Developing Countries

Globally, 56% of women have financial access as opposed to 72% of men. The gap is smaller for developing nations at 8 percentage points; however, only 67% of men and 59% of women have access (Demirgüç-Kunt, 2017). According to Demirgüç-Kunt (2017), women constitute about 56% of unbanked individuals. Unbanked individuals are typically less educated, lack documentation, and earn too little income to be able to open an account. Barriers to financial inclusion more adversely impact women than they do men, and in some developing nations, the gap in access reaches 30 percentage points (Demirgüç-Kunt, 2017). In some countries, sizeable gender gaps cause slow economic growth. The authors emphasise that any efforts to increase access must prioritise female inclusion (Demirgüç-Kunt, 2017).

In the 2020 WEF Gender Equality Report, economic participation is the second-largest gender gap at 57.8%, which has decreased from 2019. The report attributes the gap in economic participation in most countries to the lack of access to credit, collateral, and other financial services (WEF Global Gender Gap Report, 2020). Decreased access reduces the abilities of women in developing nations to start or grow businesses and eliminates opportunities to save for the future. Given the rate of change over the past 15 years, the report estimates it will take 257 years to achieve gender parity. Next to political participation, economic empowerment is a verifiable method of encouraging equality as well as economic growth in underdeveloped economies. In the realm of accessing credit 25 countries out of the 153 studied in the report do not have full inheritance rights, and 72 countries do not have the right to open bank accounts or obtain credit (WEF Global Gender Gap Report, 2020). Women in both developed and developing nations conduct the majority of unpaid domestic work. Even in nations with the lowest disparity in unpaid housework, women still complete double the amount of work as men. It is evident from this report that a combination of government policy and social-cultural norms creates inequality. However, increased financial access could be a critical method of reducing disparity.

Van Staveren (2001) analyses gender bias in finance, noting that at the micro-level men and women are comparatively different on financial behaviour. Several factors contribute to the difference in behaviour, but Van Staveren (2001) highlights the income gap as the most significant. In developing countries, the majority of women participate in unpaid work, such as household care and child-rearing, or find employment in low-skill and low-wage work. The initial factors of employment and pay give an unequal basis for financial inequality to grow further. Van Staveren (2001) finds 'gender distortions' in financial markets, which create a disadvantage for females as they also lack collateral to participate in loan activity. Overall financial literacy is low in developing nations; however, the gap between genders persists. Men are more financially literate than women purely because they have greater access and, therefore, experience in financial services. When women can actively invest and participate, they invest responsibly and have higher repayment rates on loans than men, assuming the woman retains control after receiving the loan. However, the author notes that in a study conducted on the Grameen Bank, only 37% of female borrowers retain control of their loans (Van Staveren, 2001). The study of economic behaviour is necessary to construct policies that are inclusive and encourage financial access. Empirical evidence suggests that while women might not own collateral in the form of land titles, they are more likely to repay loans, thus making them ideal candidates to increase financial services in developing nations.

Pallavi Chavan (2008) analyses the "extent and nature of gender inequality in the provision of banking services in India" (Chavan, 2008). The author uses two indicators to represent access to banking services: level of credit supplied to women and the level of deposits received by women. Chavan (2008) breaks down the data into subgroups to compare access between rural and urban areas and finds a significant disparity between the two. Women in urban areas have significantly better access to financial services than women in rural areas. Geographic location plays a vital role in a woman's ability to access financial services. Travel time, safety, and potential loss of productivity discourage women from accessing financial services, especially in rural areas. The study also found that proportionately, women contribute more deposits than they receive credit loans than men from financial institutions. Females can access financial services but are still limited in participation, highlighting the critical issue of inequality between genders. There are a few reasons men take out more loans than females: men are more willing to take risk, men make the decisions for the whole household, and men are more likely to take out loans to start businesses. Overall, Chavan (2008) points out key factors that contribute to unequal access to financial services between genders, but within females as well.

According to Seema Jayachandran (2015), unequal access to financial services and limited participation stem from more profound cultural practices that limit female autonomy to utilise services actively. The gap, caused by a lack of personal autonomy and participation in the workforce, is further exacerbated by non-existent financial independence. In Northern Africa, the Middle East, and India, social-cultural norms restrict women's independence and physical mobility, which in turn deepens unequal education, employment, and financial freedom. Traditional gender roles perpetuated by religious and cultural practices hinder the development of nations. It is essential to include the dimensions of inequality when analysing unequal access to financial services in developing nations. Country-specific characteristics create initial inequality that perpetuates unequal financial access. Gendered social constructs significantly influence the behaviours of women in developing nations. Physical mobility and independence, coupled with inequality in the eyes of the law, make accessing finance a dangerous act. Violence against women is common in developing nations and often occurs when women do not abide by social norms. If women do not feel safe accessing finance, then equal access will never occur. Some policies are attempting to circumnavigate the safety issue by providing mobile accounts, allowing women to take control of their futures without waiting for society to change first.

A report by John Issac (2014) states: "Women disproportionately face barriers to accessing finance that prevents them from participating in the economy and improving their lives" (Issac, 2014). Female-owned businesses make up around 40% of small to medium firms in developing markets; however, a majority of the financial needs are unmet by current financial services. Due to the industry that most of these businesses operate, financial access is limited, but governmental and social institutions contribute to poor access. Female-owned organisations specialise in low-skill sectors, and their ability to compete is limited. If given the same education and fair access, the most productive businesses and most innovative ideas will succeed. Policies must improve resources for women in business if economies seek to become competitive on the world stage and attain visible growth.

Klapper and Parker (2011) conduct a review on firm ownership and analyse the difference between male- and female-owned firms. Male- and female-owned firms differ between skill levels and industry types as well as gender. The majority of female entrepreneurs operate in labour-intensive, low-skill, and highly competitive sectors. Businesses in this sector rely heavily on informal financial services and require less funding than a capital-intensive high-skilled operation, especially for developing countries. However, the direction of the relationship is undetermined; however, barriers in access could be driving women to pursue business ventures in low-skilled industries leading to distortion. According to the authors, there is no explicit discrimination by the financial institutions found, but institutional and social factors create most of the barriers to entry. Klapper and Parker (2011) note there is some evidence that women did face higher interest rates for the loans, but other factors like less start-up capital provided by the entrepreneur could account for the higher interest rate. When women do receive loans for business, there is a positive effect on economic growth in the nation. The authors' analysis emphasises the need to well-rounded quantitative analysis that includes country-specific development indicators.

Inequality and Access in Sub-Saharan Africa

Blackden et al. (2007) discover gender-based barriers that impede economic growth in Sub-Saharan Africa (SSA), where poverty is the highest among developing nations. Gender inequality spans multiple dimensions, but education, employment, control of assets, and governance issues are the significant areas highlighted by Blackden et al. (2007). All dimensions significantly contribute to gender inequality in SSA; however, the authors did not include financial access. Sub-Saharan Africa experiences such sparse growth due to poor infrastructure and policy. Simultaneous improvements to education, physical infrastructure, financial access, and governance are necessary to experience meaningful growth. The freedoms and rights of women vary by country and region, but the common thread that ties SSA together is the underutilisation of resources, including human, physical, and technological capital. The mutually beneficial relationship of financial access and human development indicators allows for a compounded effect that could drive significant growth in the region.

Ambreena Manji (2010) analyses development policy aimed at increasing participation in financial services for Eastern Africa and the effect the policies have on gender inequality in the region. The author notes that previous World Bank reports emphasise reforming the financial sector to promote growth and access to financial services. The use of land titles as collateral for accessing credit and participating in financial services is one main suggestion by the World Bank. However, inheritance laws and low wages exclude women from ownership, both legally and financially. Policies seeking to increase financial access by using tangible assets, like land titles, as collateral, actively exclude women and the impoverished. Thus, this policy, and others like it, continues a pattern of institutional gender inequality in SSA and increases the adverse long-term effects of limited access. Manji (2010) argues that this biased policy perpetuates the cycle of inequality resulting from socio-cultural practices. The author raises concern over well-intended policies put forth by supranational agencies like the World Bank that lack the insight necessary to reduce negative spill-overs. Current policies choose short-term improvements at the cost of long-run equality and inclusion.

Toyin Segun Ogunleye (2017) researches microfinance and the role of women in improving participation in financial services in Nigeria. The author suggests financial inclusion as a leading solution to decreasing poverty for men and women. In the analysis, women tend to be less risky and exercise more caution than men in financial decision-making. Given this assumed behaviour, Ogunleye (2017) tests the instances of female repayment of microfinance loans. The results prove that increased microfinance loans to women in Nigeria improve the repayment rates and decrease the lender's risk. Increased microfinance for females not only reduces the gender inequality created by poor access but also encourages microfinance as an efficient method of reducing economic and gender inequalities.

Johnson and Nino-Zarazua (2011) investigate financial access and exclusion in Kenya and Uganda, emphasising country-specific research that encompasses social institutions and other unique variables. The authors find that factors such as employment, education, gender, and income heavily influence access to formal financial services. There is no difference in access between rural and urban residents. About 21% of adults in SSA have a mobile money account, which reduces the influence of geographic location to access. Johnson and Nino-Zarazua (2011) highlight the importance of using country-specific studies and condemn the one-size-fits-all policy campaigns that are usually favoured by supranational organisations. The authors also investigate the level of formal financing by breaking down the dataset analysis to informal, semi-formal, and formal, which capture significant cultural and socioeconomic factors. Empirical findings support gender inequality favouring men in accessing formal financial services; however, in semi-formal and informal sectors, access is greater for women than men. These findings prove the importance of utilising a country-specific analysis to capture certain cultural and social practices and improve accuracy. Gender inequality in financial access depends on the nation and its unique characteristics.

Aterido et al. (2013) investigate the differences in financial services accessed by households, breaking down their analysis into equality of financial services between personal accounts and business accounts. Access to finance for business includes formal and informal financing and the difference between genders in accessing formal financing channels. Initially, findings suggest that there is a gender gap; however, once controlled for external characteristics the authors find no difference between male and female access of formal financing for business (Aterido et al., 2013). Firm characteristics appear to create the inequalities present, especially size. Selection bias contributes to observed inequality, as female entrepreneurs have higher barriers to entry than their male counterparts. The authors find a definitive unconditional gap in access; however, there is no inequality between genders at the household level in the use of formal services once controlled for other characteristics. Aterido et al. (2013) suggest that financial services themselves do not create inequality; instead, external characteristics influence the ability to access formal financial services. Education, income level, employment, and age are all key characteristics that affect accessing formal financial services.

Asiedu et al. (2013) note that current empirical findings suggest a lack of financial access is the primary constraint for all firm growth in Sub-Saharan Africa. Private enterprises remain underdeveloped due to unmet funding and economic instability. Female-owned firms face more constraints than male-owned firms, and SSA firms are more constrained than any other region. Thus, female firms in SSA encounter the most adversity in financing. Asiedu et al. (2013) found that SSA female-owned firms experience 5.2% more in constraints than male-owned firms. After controlling for other characteristics and bias, the gender gap persisted, highlighting the severe disparity in financial access.

Hansen and Rand (2014) test the accuracy of survey data in determining unequal credit access of SSA firms by gender. Self-reported survey data measures the perceived inequality between male- and female-owned businesses in accessing credit. The authors find a marginally significant gap between firms when using self-reported data (Hansen & Rand, 2014). However, when using formally reported data from financial institutions, no gender gap is found between male- and female-owned firms. Hansen and Rand's (2014) research addresses discrepancies in measuring gender inequality in business, but access data is limited, and surveys are the best method of capturing actual and perceived access. Self-reported measures shed light on how people view access to finance and capture information on education, age, income, location, and other qualitative factors. The discrepancy between survey data and institutional data proves the influence of institutional factors on access and refocuses the goal of increasing access back to reducing gender equality in all dimensions.

Nwosu and Orji (2017) analyse the impact of gender and formal credit access on small- to medium-sized firm performance in Nigeria. Credit constraints reduce productivity and reinvestment into SMEs. The effect of credit constraints becomes more significant for female firms than male firms once adjusted for estimator biases. Formal access to credit has a significant impact on firm performance. Informal access to credit does provide a finance option that would otherwise not exist; it is not a reliable method of increasing firm growth and productivity. Equal access to formal financial services allows the most productive and competitive firms to grow. Nwosu and Orji (2017) suggest that policy reforms are necessary to increase access to formal credit loans and that will, in turn, improve performance of SMEs, especially female-owned firms in Nigeria.

The existing literature fails to encompass all three components, gender inequality, financial access, and development, into one research question. Research that does incorporate at least two of the factors uses two methods of analysis: statistical analysis and country-specific policy analysis. Much of the literature regarding gender equality and access mainly analyses the disparity using literature reviews without statistical analysis or analyses multiple countries and omits relevant country policies and characteristics. Quantitative analysis is compulsory in development as policies are created surrounding informed decision-making. Therefore, to ensure useful policy, research must combine an understanding of the nation in question with the necessary statistical analysis to determine the best path forward. Numerous factors influence gender equality and financial access, thus requiring the correct variables and detailed models to obtain significant results.

Financial Inclusion in Rwanda

The importance of financial access in achieving gender parity is evident from the previously reviewed literature. Rwanda is no exception, and the case for equal access is bolstered by the nation's high global parity ranking and low economic opportunity rankings. Rwanda-specific literature analysing gender equality and financial access is limited, thus, leaving much of the topic unexplored. Rwanda's gender parity performance is impressive, but financial access needs improvement for the nation to experience significant economic development.

According to the WEF Global Gender Gap Report (2020), Rwanda ranks first in Sub-Saharan Africa and ninth globally in gender parity. The nation experiences 79.1% gender parity, but equality varies across measures of equality. Political empowerment in Rwanda ranks in the top four in the world, with over 50% of parliamentarians and ministers being female (WEF Global Gender Gap Report, 2020). Educational attainment is equal across primary and secondary enrolment; however, only 69.4% of women and 77.5% of men are literate. Overall tertiary enrolment is low at 8%. However, men are twice as likely to pursue further education as women (WEF Global Gender Gap Report, 2020). The shortcoming in education translates to the lowest parity indicator, economic participation and opportunity, which rests at 67.2%. Men and women participate equally in the labour force, but disparities arise in wages, professional employment, and leadership. A combination of unequal higher education enrolment and socio-cultural norms all contribute to economic inequalities. Rwanda is the first in the world with perfect parity in labour force participation. The WEF Global Gender Report (2020) excludes financial access measures from parity calculations, potentially missing out on essential indicators of equality. The report suggests investments to education and human capital as crucial methods to increase parity. While improvements to these areas are undeniable, financial access is proven to influence gender inequality; therefore, Rwanda's gender parity analysis is incomplete without it.

Tom Goodfellow (2017) explains the unique development nature of Rwanda by labelling the nation as 'neo-developmental'. The author suggests that Rwanda and other African countries underutilise taxes and difficulty in attracting investment. A lack of capital as a result of land leasing and tax policy severely impacts Rwanda's ability to grow business and manufacturing. Goodfellow (2017) acknowledges that development occurs sequentially and simultaneously but highlights the necessity of economic growth. While the author highlights specific policy regarding land and taxes, financial access is again excluded from the analysis. Economic development is an ecosystem where economic growth relies on the wellbeing of all entities like education, financial access, infrastructure, and others. Goodfellow (2017) proposes increased property taxes as a method of increasing capital but overvalues its contribution to quality human and economic development.

Mukamana et al. (2016) evaluated Duterimbere microfinance and its role in increasing equal access to credit loans in Rwanda. In Rwanda, only one out of five women holds an account at a formal financial institution as opposed to four out of five men (Mukamana et al., 2016). The authors suggest the inequality in access arises from discriminatory policies against women. The main barrier to obtaining a credit loan is the requirement of collateral. The Rwandan government passed numerous initiatives, like Vision 2020, to increase access and reduce inequalities between genders. While tailor-made programmes meant to increase equality, the benefit of the initiatives is unknown. Mukamana et al. (2016) conducted a series of questionnaires to determine loan amounts and the frequency of accessing credit among women in Rwanda. The results reveal that microfinance initiatives improved equality in accessing loans through the tailoring requirements to increase the accessibility of credit. Eliminating collateral requirements and introducing group loans made access more widely available for women in Rwanda.

As noted by the WEF Global Gender Report (2020), Rwanda's educational enrolment is relatively equal across genders. Comparing primary and secondary gross percentage enrolment rates across genders in Fig. 4.1 reveals that primary enrolment rates are relatively steady over time. However, there is a stark contrast between primary and secondary enrolment rates for both genders. While secondary enrolment increased since 2010, rates remain significantly lower than primary enrolment. The key take-away from Fig. 4.1 is the difference in enrolment between genders. Both male and female access remain relatively equal over time, with most years experiencing higher female enrolment rates than male enrolment rates in both educational levels.

Another critical aspect of gender equality in Rwanda is literacy. The 1994 genocide interrupted education from 1994 onwards until the late 1990s. The literacy rate captures adults over the age of 15 who can read and write, thus including the children affected by the genocide. Figure 4.2 compares the literacy rates of the adult population in Rwanda from 2010

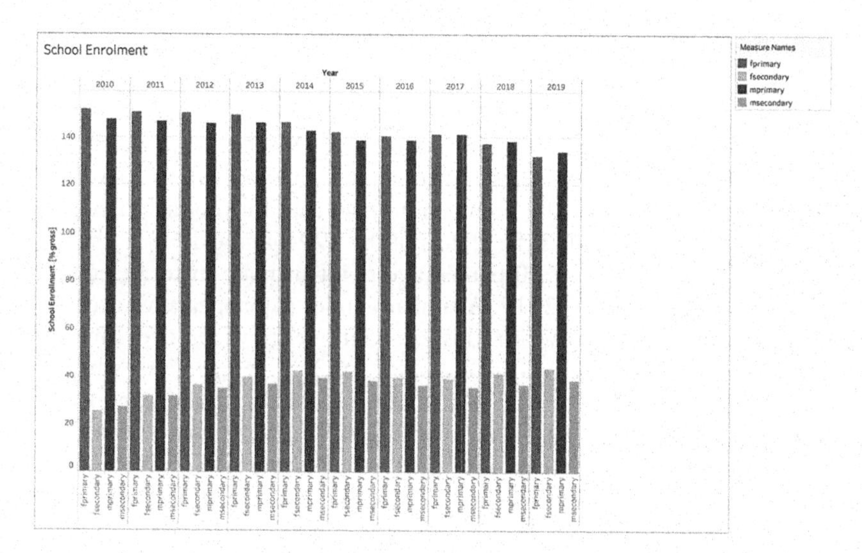

Fig. 4.1 School enrolment

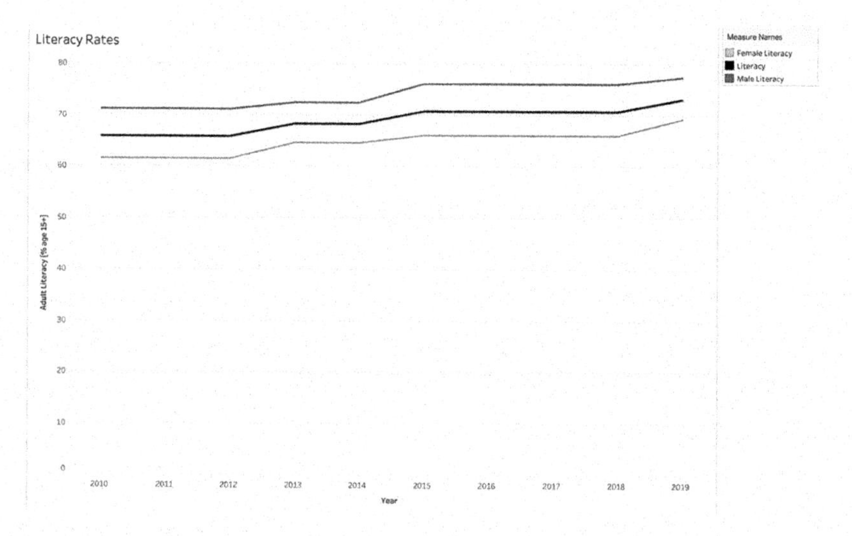

Fig. 4.2 Literacy rates

through 2019. There is a consistent 10-percentage point gap between male and female literacy rates with women experiencing lower literacy. While the total literacy rate increased from 65% to 73%, women are consistently less literate than men. Figure 4.2 suggests that gross enrolment rates are equal across the sexes; however, the difference in literacy casts doubt on this equality. The literacy rate may capture residual effects of gender inequality as well as the effects of the 1994 genocide.

Figure 4.3 illustrates a 20-percentage point increase in total account ownership from 2010 to 2019. A similar gender gap appears for account ownership in Rwanda; however, this gap continues to widen over time. In 2010 there is only a five-percentage point gap between genders, whereas in 2015 the gap widened to 15 percentage points before shrinking to 10 percentage points in 2019. While total access is increasing over time, it is apparent that the increase is not equal across genders. For Rwanda to experience the long-term benefits of increased financial access, men and women should, at the very least, experience equal growth rates.

Finally, Fig. 4.4 compares the type of accounts owned by males and females in Rwanda. A similar gap persists over time between the sexes for financial institution accounts. In 2015 total and male-owned financial institution accounts increased to 45% but decreased to 40% in 2019. Over

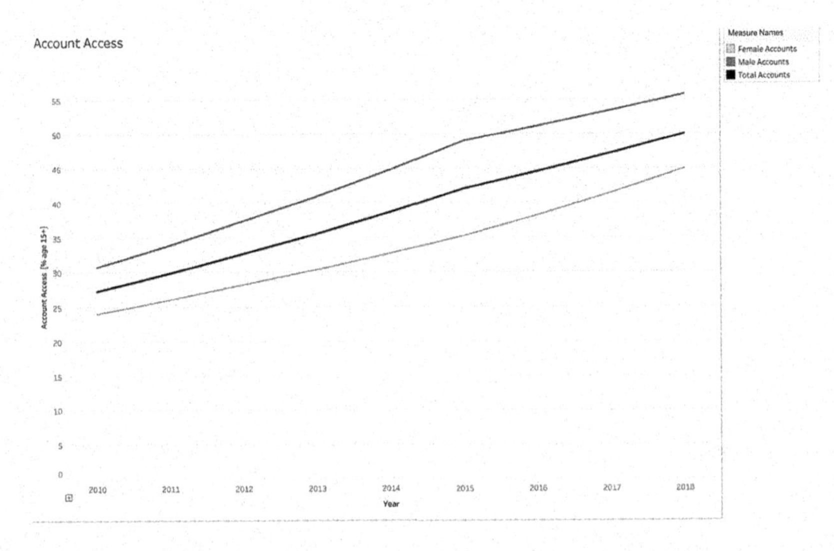

Fig. 4.3 Access to bank accounts

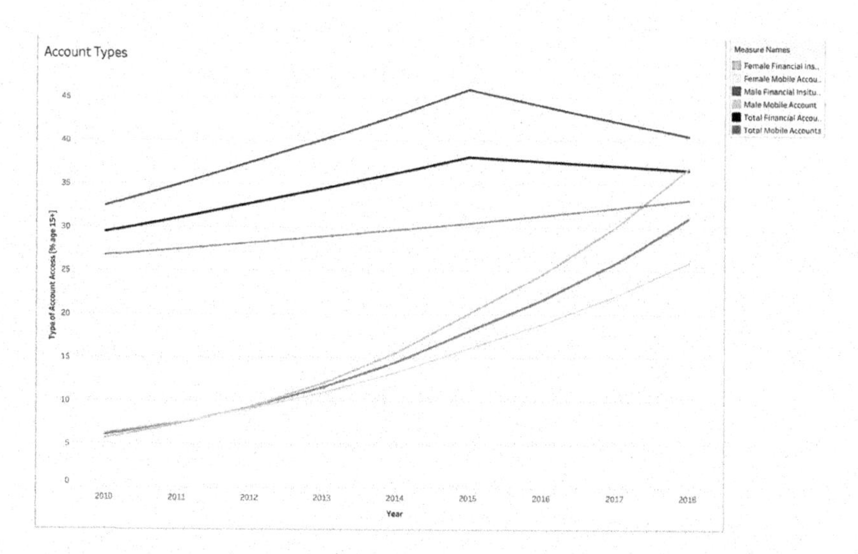

Fig. 4.4 Bank account types

the same period, mobile money accounts increased by 15 percentage points across both males and females. Increased access to mobile phones and the internet also created a new method of account ownership in Rwanda. Should mobile money accounts continue this trend, mobile accounts would surpass financial institution accounts in the next few years. Mobile banking is a convenient method of storing, sending, and receiving money. Recent initiatives promote mobile money accounts as the most efficient method of increasing financial access in developing nations.

Gender gaps exist across numerous dimensions; however, the gap in financial access is the most persistent. School enrolment rates suggest that younger generations might not experience the same gaps as older generations. However, disparities in literacy rates, wages, and employment skill levels suggest that equality remains out of reach. Equal account access would ensure all people have a safe place to receive wages, save for the future, and actively participate in the financial economy.

Given the empirical findings of previous research, it is evident that gender equality and financial access are necessary for economic development. The one-size-fits-all policies and checklists implemented in the postcolonial era proved ineffective, creating the opportunity to research and utilise new data-backed methods of development. The importance of human development indicators and infrastructure remains; however, the influence of gender equality and financial access on economic growth is now considered the leading determinant of economic development. It is now possible to conduct statistical analysis on these two factors due to the data collection efforts of NGOs over the past ten years. Country-specific research and custom-made policies are now possible with newly available data, furthering the significance and the magnitude of popular measures of development.

Numerous growth theories and empirical findings support the linkages between equality, access, and development. An equally and fairly accessible financial system plays a vital role in economic development. Development in itself is an aggregate measure of multiple indicators acting simultaneously to either positively or negatively impact growth. It is difficult to pinpoint the directional relationship of development and indicators, like education, financial access, wages, and employment. Therefore, quantitative analysis is conducted on a case-by-case basis to capture individual country characteristics. According to Arestis and Demetriades (1997), in half of the countries analysed economic growth leads to financial development; however, this relationship does not apply to all nations, and financial access can spur economic growth. The authors recommend utilising time-series data and country-specific measures to increase accuracy within the models. Traditionally, GDP per capita acts as the indicator of economic development, and its use in this instance is supported by Arestis and Demetriades (1997) as the best measure.

Johnson and Zarazua include social institutions to determine how they interact with financial access and equality. The authors suggest the use of proxy measures like education, wages, number of accounts, and others to quantify social institutions. Given the nature of the measures, it is necessary to analyse the variables on a national level as Johnson and Zarazua have done. Just in their research the difference in results between Kenya and Uganda exemplifies the reasons for utilising country-specific data. Arestis and Demetriades (1997) and Johnson and Zarazua provide helpful insight that shapes the methodology utilised in the forthcoming analysis on financial access and equality in Rwanda that employs times-series data between 2011 and 2019.

Demirgüç-Kunt (2017) measure individual financial access for over 140 economies using the Global Findex database. The Global Findex database includes numerous indicators of formal and informal financial access. The authors clearly define each variable utilised in determining multiple financial access dimensions (Demirgüç-Kunt, 2017). Account access is measured using the percentage of adults owning accounts, which is specified further by gender. There is a notable gap between genders in account access, but the authors do not conduct the statistical analysis to determine the impact of the gap on development. The research conducted on Rwanda seeks to determine the significance and magnitude of this relationship using OLS regression analysis. Demirgüç-Kunt (2017) clearly define the most critical measures in understanding financial access. The models determining access in Rwanda include critical indicators like the percentage of adults with accounts, savings, and mobile money accounts.

The literature review emphasises the importance of multiple factors like income, education, and the ability to provide collateral in determining access. These are the same factors that influence gender equality in general, as well as access to finance. The WEF Global Gender Report (2020) and Sarma et al. highlight the importance of gender equality on access and, in turn, economic development. Human development indicators play a massive role in understanding the quality of equality and access. The WEF Gender Report (2020) includes an index of equality variables across health, education, inclusion, and political empowerment. Equality and development indicators represent the qualitative nature of gender equality, especially in Rwanda. The nation appears to be equal, ranking ninth in the world; however, some essential areas like wages and unpaid household duties suggest inequalities in income and gender roles. This research includes the influence of critical indicators like education, income, employment, and representation to determine their significance and impact on gender equality, access, and economic development in Rwanda.

The previous research literature provides the framework for variable selection, as well as the best databases to source such data. Demirgüç-Kunt (2017) and the WEF Global Gender Report (2020) provide the sources of data and from their references and data for this analysis is sourced from the following:

- World Bank Human Development Indicators database
- World Bank Global Financial Inclusion database
- International Labour Organisation Labour Statistics database

The data can be split into two major categories as inspired by the WEF Global Gender Report (2020) and Demirgüç-Kunt (2017). The first attempts to capture the dimensions of gender inequality in Rwanda. The WEF Global Gender Report (2020) highlights education, literacy, and parliamentary representation as significant inequality indicators. The second category includes specific measures surrounding financial access. Demirgüç-Kunt (2017) evaluate the entirety of household account usage, detailing how households paid bills, received remittances, and if they paid bills online. For this research account access, financial institution accounts, and mobile money accounts capture the critical indicators of access in Rwanda. Table 4.1 provides the details of each variable, including their sources and definitions.

The majority of variables selected are available on an annual basis; however, financial access data occur less regularly. The Global Findex database began recording data in 2012 from Rwandan financial access surveys. The World Bank re-evaluated access levels in 2015 and 2018. For this research, consistent yearly data is required to complete the time-series ordinary least squares regressions and determine the significance of each selected variable on economic growth. Therefore, the annual percentage change is calculated for each missing year using the two most recent years of data. Each missing value is an approximation of what the values would most likely be, given the percent change over the period. For example, for a given period between 2012 and 2015, the annual percentage change is 5%. Using the 2012 value, the equation would be as follows.

$$\mathbf{2013\ value = (2012\ value \times 0.05) + 2012\ value}$$

Without this approximation of values, it would be impossible to conduct an OLS regression. Research on financial access and its influence on development is limited, and the necessary data is even more sparse among developing nations like Africa. However, this only reinforces the need for more analysis on the subject, even if estimations must be employed.

Arestis and Demetriades (1997) and Johnson and Zarazua recommend using country-specific times-series data to capture the individual effects of financial access and gender equality. Using the variables explained above, three time-series regressions are employed using the ordinary least squares method. The first regression aims to determine the influence and significance of societal gender equality on female financial access. The natural log of female account ownership represents the percentage growth in

Table 4.1 Regression variables

Variable	*Abbreviation*	*Definition*	*Source*
GDP per capita (current US$)	gdpcapita	GDP per capita is gross domestic product divided by midyear population	World Bank national accounts data
Account ownership (% age 15+)	account faccount maccount	Percentage of population (total, female, male) aged 15+ who report having an account (by themselves or together with someone else) at a bank or mobile money service in the past 12 months	Global Findex database
Mobile money account ownership (% age 15+)	mobileacct fmobileacct mmobileacct	Percentage of respondents who report personally using a mobile money service in the past 12 months	Global Findex database
Financial institution account (% age 15+)	finacct ffinacct mfinacct	Percentage of respondents who report having an account (by themselves or together with someone else) at a bank or another type of financial institution	Global Findex database
School enrolment, primary (% gross)	primary fprimary mprimary	Gross enrolment ratio is the ratio of total enrolment, regardless of age, to the population of the age group that officially corresponds to the level of education shown. Primary education provides children with basic reading, writing, and mathematics skills along with an elementary understanding of such subjects as history, geography, natural science, social science, art, and music.	UNESCO Institute for Statistics
School enrolment, secondary (% gross)	secondary fsecondary msecondary	Gross enrolment ratio is the ratio of total enrolment, regardless of age, to the population of the age group that officially corresponds to the level of education shown. Secondary education completes the provision of basic education that began at the primary level and aims at laying the foundations for lifelong learning and human development, by offering more subject- or skill-oriented instruction using more specialised teachers	UNESCO Institute for Statistics

(*continued*)

Table 4.1 (continued)

Variable	*Abbreviation*	*Definition*	*Source*
Literacy rate, adult (% of people ages 15 and above)	literacy fliteracy mliteracy	Adult literacy rate is the percentage of people aged 15 and above who can both read and write with understanding a short, simple statement about their everyday life	UNESCO Institute for Statistics
Proportion of seats held by women in national parliaments (%)	fparliament	Women in parliaments are the percentage of parliamentary seats in a single or lower chamber held by women	Inter-parliamentary Union
CPIA gender equality rating (1=low to 6=high)	cpia	Gender equality assesses the extent to which the country has installed institutions and programmes to enforce laws and policies that promote equal access for men and women in education, health, the economy, and protection under law	World Bank Group

female account ownership and reduces potential effects of heteroscedasticity given the rapid increase in accounts over the ten-year period.

$$\text{lnfaccount} = \alpha + \beta_1 \text{fliteracy} + \beta_2 \text{fparliament} + \beta_3 \text{fprimary} + \beta_4 \text{fsecondary} + \mu$$

H_0: Gender equality measures do not influence female financial access in Rwanda.
H_1: Gender equality measures do influence female financial access in Rwanda.

According to Johnson and Zarazua, the WEF Global Gender Report (2020), and Sarma et al., it is predicted that all gender equality measures will positively and significantly impact female financial access.

The second regression attempts to determine the influence and significance of female financial access dimensions on overall economic growth.

$$\text{lngdpcapita} = \alpha + \beta_1 \text{fmobileacct} + \beta_2 \text{ffinacct} + \mu$$

H_0: Female financial access does not influence GDP per capita *in Rwanda.*

H_1: Female financial access does influence GDP per capita *in Rwanda.*

Given the research and results of Demirgüç-Kunt (2017), it is projected that all dimensions of female financial access will positively and significantly impact economic growth. Any increase in access is assumed to impact all areas of the economy positively, thus positively impacting GDP per capita. Using the natural log of GDP per capita captures, the percent growth and accurately measures increases in development for Rwanda. If access increases over time, infrastructure, human capital, and financial wellbeing will improve as well. By increasing access, consumers begin thinking about finances in the long term as opposed to the short term.

The third and final regression seeks to understand the relationship and significance of total female account access and Rwanda's CPIA Gender Equality rating on GDP per capita. This model combines the equality measures utilised in the first regression with the access measures employed in the second regression to understand their aggregate benefits on GDP per capita.

$$\text{lngdpcapita} = \alpha + \beta_1 \text{faccount} + \beta_2 \text{cpia} + \mu$$

H_0: Female financial access and gender equality do not influence GDP per capita *in Rwanda.*
H_1: Female financial access and gender equality do influence GDP per capita *in Rwanda.*

Combining the findings of Demirgüç-Kunt (2017), Johnson and Zarazua, the WEF Global Gender Report (2020), and Sarma et al. both female financial access and gender equality should positively and significantly impact GDP per capita in Rwanda.

Gender gaps persist in numerous areas of Rwanda economy and society. The gap in financial access is apparent, and without an equal increase in account ownership, sustainable long-term growth is not a reality. After carefully selecting appropriate variables using the guidance of Demirgüç-Kunt (2017) and the WEF Global Gender Report (2020), three models are constructed using the methods suggested by both Arestis and Demetriades (1997) and Johnson and Zarazua, thus resulting in three times-series linear models that employ ordinary least squares to determine the relationship and significance of gender equality, financial access, and economic growth in Rwanda. The following chapter discusses the results of the statistical analysis and how the findings relate to the current literature on gender equality, financial access, and economic development.

The linkages between financial access and gender equality are apparent; however, how they interact and whether or not it is significant to economic development in Rwanda remains undetermined. Reports by Demirgüç-Kunt (2017) and the WEF Global Gender Report (2020) provide important statistics surrounding access and gender equality but do not utilise econometric analysis to determine the relationship between these variables. In employing econometric analysis this research seeks to understand the relationship between gender equality, financial access, and economic growth over time in Rwanda. This research also hopes to capture the effects of recent initiatives by the World Bank and Rwandan government to increase financial access and gender equality by using the data from 2010 through 2019.

Regression 1: The Effect of Gender Equality Dimensions on Female Financial Access

As explained previously Regression 1 attempts to determine the relationship between gender equality measures and female account ownership in Rwanda. Table 4.2 includes the results of the OLS regression analysis for Regression 1. In general regression, one does not support the connection between gender equality dimensions and female account ownership in Rwanda. Female literacy, parliamentary representation, and secondary enrolment rates are insignificant at all levels. The control variable is significant at a 95% confidence interval, suggesting that a one-unit increase in other factors accounts for a 216,408.5215% increase in female account ownership. Therefore, numerous other dimensions significantly impact female financial access in Rwanda.

Female primary school enrolment is significant at a 95% confidence interval. However, there is a negative relationship between primary enrolment and female account ownership, suggesting that increasing female primary enrolment by one unit results in a 3.32% decrease in female account ownership. Previous literature suggests a positive relationship between school enrolment and account ownership, but there are a couple of reasons why the relationship is negative here. Gross female primary enrolment rates are available data concerning school enrolment from 2011

Table 4.2 Regression 1 results

Dependent variable: lnfaccount	
Control	7.680215**
	(2.655109)
fliteracy	0.013519
	(0.027034)
fparliament	-0.007154
	(0.010149)
fprimary	-0.033771**
	(0.008838)
fsecondary	0.007474
	(0.006140)
R^2	0.963865
Adjusted R^2	0.927729
Observations	9 after adjustments
Standard errors in parentheses	
*** significant at 1%	
** significant at 5%	
* significant at 10%	

to 2019. Since the gross value includes students of all ages, adults actively enrolled in education are less likely to own an account. Another reason for the unexpected relationship is the long-term influence of education. Given that education enrolment accounts for those enrolled in school now, this might not accurately represent the influence of education on account ownership. Educational attainment for the female population ages fifteen and older could better represent the population of adult account holders.

The model is free of both serial correlation and heteroskedasticity, supporting the viability of the model. The regression results suggest that education, political representation, and literacy rates do not impact female account ownership in Rwanda, as suggested by the WEF Global Gender Report (2020). The methodology suggested by Arestis and Demetriades (1997) and Johnson and Zarazua and employed in this analysis proves that country-specific research is vital to understanding the real impact of a nation's policy. Future research requires data that accurately depicts the current adult population. As is usually the issue in researching developing nations, data availability hinders researchers' ability to fully understand the country and what policies would best increase financial access and economic development.

Regression 2: The Effect of Female Account Types on GDP per Capita in Rwanda

The second regression attempts to understand what types of female accounts influence GDP per capita in Rwanda. There are two main methods of owning an account in Rwanda: mobile accounts and financial institution accounts. While financial institution accounts are the traditional method of financial access, mobile account ownership is growing in popularity as a method to increase account ownership in developing nations (Demirgüç-Kunt, 2017).

Table 4.3 contains the results of Regression 2 and suggests that female mobile money accounts are significant at all levels. However, the relationship between female-owned mobile accounts and GDP per capita is negative. A one-unit increase in female mobile accounts results in a 5.71% decrease in GDP per capita. Mobile account usage differs greatly from that of a financial institution account. Mobile accounts leave owners open to identity theft, data theft, and other scams (Poverty and Privacy, 2020). While this might not be the reason for the negative relationship, it does highlight the dangers of mobile accounts in developing nations with naïve populations.

The relationship between female-owned financial institution accounts and GDP per capita is both significant at all levels and positive. One unit increase in female-owned financial institution accounts results in a 29.57%

Table 4.3 Regression 2 results

Dependent variable: lngdpcapita	
Control	15.83738***
	(1.015416)
fmobileacct	-0.058831***
	(0.012879
ffinacct	0.259024***
	(0.040227)
R^2	0.981638
Adjusted R^2	0.975530
Observations	9 after adjustments
Standard errors in parentheses	
*** significant at 1%	
** significant at 5%	
* significant at 10%	

increase in GDP per capita for Rwanda. Therefore, Rwandan policies should focus on increasing female account ownership at financial institutions to increase GDP per capita. The influence of the control on GDP per capita suggests that numerous other dimensions significantly impact GDP per capita. There is no evidence of serial correlation or heteroskedasticity, thus proving the reliability of the model. While Demirgüç-Kunt (2017) champion the usage of mobile accounts for increasing financial access, the negative impact on GDP per capita makes a case for financial institutions and their role in Rwandan economic development. The results also provide support for including female financial access as a determinate of development and equality. The WEF Global Gender Report (2020) excludes financial access measures from calculating gender parity scores. However, given the influence of female account ownership in Rwanda, the inclusion of financial access would better represent equality and the contribution of women to economic development.

Regression 3: The Effect of Female Account Ownership and Gender Equality Rating on GDP per Capita in Rwanda

Regression 3 combines the influence of gender equality and female financial access to determine if the two dimensions significantly influence GDP per capita. Demirgüç-Kunt (2017), Johnson and Zarazua, the WEF Global Gender Report (2020), and Sarma et al. all suggest that financial access and gender equality are essential for increasing economic development. Table 4.4 suggests that only female account ownership significantly impacts GDP per capita in Rwanda. Female financial access is significant at a 95% confidence interval, and a one-unit increase results in a 1.72% increase in GDP per capita. The CPIA gender equality rating is insignificant at all levels, thus suggesting that the institutions and programmes utilised in calculating the rating do not impact GDP per capita. The control is significant at a 99% confidence interval, suggesting that a one-unit increase in other factors is responsible for 60,778.06% increase in GDP per capita. Total female account ownership proves to impact GDP per capita significantly. The model initially suggested some heteroskedasticity, but after conducting a White test, no heteroskedasticity exists. The model is also free of serial correlation, further proving the efficiency of the model. Single-country analysis using time-series data as suggested by Arestis and Demetriades (1997) and Johnson and Zarazua proves that each nation is different and must tailor their policies to fit their unique situation.

Table 4.4 Regression 3 results

Dependent variable: lngdpcapita	
Control	6.411458***
	(0.491785)
faccount	0.017056**
	(0.005693)
cpia	-0.109794
	(0.153435)
R^2	0.7995904
Adjusted R^2	0.727871
Observations	9 after adjustments
Standard errors in parentheses	
*** significant at 1%	
** significant at 5%	
* significant at 10%	

The notion that gender equality does not impact economic development contradicts the works of Johnson and Zarazua and Sarma et al. However, it supports the findings of Demirgüç-Kunt (2017), Asiedu et al. (2013), Ogunleye (2017), Beck et al. (2009), and Arestis and Demetriades (1997) in that financial access does in fact influence development. Female access does indeed contribute to economic development, but nations must understand what policies increase financial access. While education, parliamentary representation, equal wages, and labour force participation remain important methods to increase overall gender equality, they do not directly increase female account ownership in Rwanda. Continued efforts are necessary to continue increasing gender equality and financial access in Rwanda.

In order for future studies to obtain even more useful results, data collection efforts by the World Bank and other organisations must continue. The main shortcoming of this research is the lack of available data. While it is necessary for consistent data, the type of data required for this type of analysis differs from other studies using human development and financial access indicators. Current data on the adult population, including educational attainment, labour skill level, and other adult level measurements, would significantly increase the accuracy of the models. Continued data collection is vital for research and policy development in developing economies. If policymakers understand what factors effectively increase development, then targeted policies will reduce the wastage of time and resources.

Conclusion

The world and its developing economies are beginning to understand the role of financial access in economic development. As the world's wealthier nations continue their upward development, developing nations struggle to keep pace. Increased financial access, especially female access, is proving itself as an effective method for developing economies to jump-start and sustain economic development. Arestis and Demetriades (1997) highlight the increasing need for well-developed financial markets and participation in developing nations. The need for individual country analysis is further emphasised by Johnson and Zarazua. These authors reject the one-size-fits-all methods of development used throughout the twentieth century and instead acknowledge the uniqueness of individual economies. Education, labour markets, infrastructure, politics, and many other factors create unique economic climates that require specific development policies.

The WEF Global Gender Report (2020) and Demirgüç-Kunt (2017) continue efforts to understand gender equality and financial access in developing nations. The reports provide an understanding of each nation's development and analyse the most critical dimensions of quality economic growth. While the results of this research do not support the linkages between gender equality and financial access, it does support the significant and positive influence of female-owned accounts. Financial access provides an opportunity for consumers to save money, securely receive wages, and obtain loans for economic prosperity. As the global economy continues to shift, the importance of financial access increases.

In conclusion, Rwanda appears to rank ninth in the world in gender parity. However, the results of this research suggest this gender parity does little to increase female economic participation in the nation. The disconnect between high gender parity and poor economic participation among women cast doubt on the importance of traditional gender equality measures. Instead, areas like financial access appear to bring about real economic prosperity for women. When women can actively save, invest, and obtain loans, they gain an element of independence not found previously. Increasing the financial independence of women in developing nations also increases their ability to advocate for equal pay, education, and opportunity. Increasing female financial access benefits not only women but the entire nation, as it is evident that GDP per capita increases with increased financial access. The key to successful economic development relies on the engagement of every individual in the economy.

References

Ang, J. B. (2010). Finance and Inequality: The Case of India. *Southern Economic Journal, 76*, 738–761. https://doi.org/10.4284/sej.2010.76.3.738

Arestis, P., & Demetriades, P. (1997). Financial Development and Economic Growth: Assessing the Evidence*. *The Economic Journal, 107*, 783–799. https://doi.org/10.1111/j.1468-0297.1997.tb00043.x

Asiedu, E., Kalonda-Kanyama, I., Ndikumana, L., & Nti-Addae, A. (2013). Access to Credit by Firms in Sub-Saharan Africa: How Relevant Is Gender? *American Economic Review, 103*, 293–297. https://doi.org/10.1257/aer.103.3.293

Aterido, R., Beck, T., & Iacovone, L. (2013). Access to Finance in Sub-Saharan Africa: Is There a Gender Gap? *World Development, 47*, 102–120. https://doi.org/10.1016/j.worlddev.2013.02.013

Batuo, M. E., Guidi, F., & Mlambo, K. (2010). Financial Development and Income Inequality: Evidence from African Countries 28.

Beck, T., Demirguc-Kunt, A., & Honohan, P. (2009). Access to Financial Services: Measurement, Impact, and Policies. *The World Bank Research Observer, 24*, 119–145. https://doi.org/10.1093/wbro/lkn008

Blackden, M., Canagarajah, S., Klasen, S., & Lawson, D. (2007). Gender and Growth in Sub-Saharan Africa: Issues and Evidence. In G. Mavrotas & A. Shorrocks (Eds.), *Advancing Development* (pp. 349–370). Palgrave Macmillan UK. https://doi.org/10.1057/9780230801462_19

Chavan, P. (2008). Gender Inequality in Banking Services. *Economic and Political Weekly, 43*, 18–21.

Demirgüç-Kunt, A. (2017). The Global Findex Database 2017 151. https://globalfindex.worldbank.org/

Goodfellow, T. (2017). Taxing Property in a Neo-Developmental State: The Politics of Urban Land Value Capture in Rwanda and Ethiopia. *African Affairs, 116*, 549–572. https://doi.org/10.1093/afraf/adx020

Hansen, H., & Rand, J. (2014). Estimates of Gender Differences in Firm's Access to Credit in Sub-Saharan Africa. *Economics Letters, 123*, 374–377. https://doi.org/10.1016/j.econlet.2014.04.001

Issac, J. (2014). Expanding Women's Access to Financial Services [WWW Document]. World Bank. Accessed July 6, 2019, from http://projects-beta.worldbank.org/en/results/2013/04/01/banking-on-women-extending-womens-access-to-financial-services.

Jayachandran, S. (2015). The Roots of Gender Inequality in Developing Countries. *Annu. Rev. Econ., 7*, 63–88. https://doi.org/10.1146/annurev-economics-080614-115404

Johnson, S., & Nino-Zarazua, M. (2011). Financial Access and Exclusion in Kenya and Uganda. *Journal of Development Studies, 47*, 475–496. https://doi.org/10.1080/00220388.2010.492857

Klapper, L. F., & Parker, S. C. (2011). Gender and the Business Environment for New Firm Creation. *The World Bank Research Observer, 26*, 237–257. https://doi.org/10.1093/wbro/lkp032

Manji, A. (2010). Eliminating Poverty? 'Financial Inclusion', Access to Land, and Gender Equality in International Development: Eliminating Poverty? *The Modern Law Review, 73*, 985–1004. https://doi.org/10.1111/j.1468-2230.2010.00827.x

Mukamana, L., Sengendo, M., & Okiria, E. (2016). Promoting Gender Equality in Access to Microcredit Through Flexible Lending Approaches of Female Targeting MFIs: Evidence from Duterimbere MFI of Rwanda. *International Journal of Business & Economic Development, 4*(3), 33–44.

Nwosu, E. O., & Orji, A. (2017). Addressing Poverty and Gender Inequality through Access to Formal Credit and Enhanced Enterprise Performance in Nigeria: An Empirical Investigation: Addressing Poverty and Gender Inequality. *African Development Review, 29*, 56–72. https://doi.org/10.1111/1467-8268.12233

Ogunleye, T. S. (2017). Financial Inclusion and the Role of Women in Nigeria: Financial Inclusion. *African Development Review, 29*, 249–258. https://doi.org/10.1111/1467-8268.12254

Poverty and Privacy: How Digital Financial Services can Prey Upon the Poor, 2020. The Economist.

Van Staveren, I. (2001). Gender Biases in Finance. *Gender and Development, 9*, 9–17. https://doi.org/10.1080/13552070127734

World Economic Forum Global Gender Gap Report 2020 (Insight Report), n.d. World Economic Forum. http://www3.weforum.org/docs/WEF_GGGR_2020.pdf.

CHAPTER 5

Is there a Gender Gap in Accessing Finance in Rwanda?

Madelin O'Toole and Bhabani Shankar Nayak

Introduction

This chapter explores the relationship between gender equality and financial access, as well as their relationship with economic development in Rwanda. Current literature on the topic suggests improvements to gender equality and financial inclusion are critical to creating sustainable economic growth in developing nations. While Rwanda ranks ninth in the world in gender parity, this chapter seeks to understand if this ranking reflects equality in financial access and economic development that can transform the lives of individuals. Both the WEF Gender Equality Report (2020) and (Demirgüç-Kunt, 2017) highlight financial inclusion as a method to reduce inequality and create long-term development in developing nations. Much of the current research including Pallavi Chavan

M. O'Toole (✉)
Federal Research Division, Library of Congress, Washington, DC, USA

B. S. Nayak
University for the Creative Arts, Epsom, UK
e-mail: bhabani.nayak@uca.ac.uk

B. S. Nayak (ed.), *Political Economy of Gender and Development in Africa*, https://doi.org/10.1007/978-3-031-18829-9_5

(2008), John Issac (2014), Seema Jayachandran (2015), van Staveren (2001), Blackden et al. (2007), and others suggest that already present inequalities in education, healthcare, labour, and wages create inequality in financial access. Asiedu et al. (2013), Hansen and Rand (2014), Nwosu and Orji (2017), and Klapper and Parker (2011) find similar results when investigating female-owned or managed firms. From the current empirical findings, socio-cultural practices that create inequality are what cause unequal financial access. Much of the literature surrounding female financial inclusion and development analyses multiple countries; however, Arestis and Demetriades (1997) and Johnson and Nino-Zarazua (2011) emphasises the importance of single country analysis over time.

Gender Inequality in Access to Finance

Modern development relies on the usage of financial inclusion to reduce income inequality and sustain steady growth. While equal access between income levels is a critical element, the influence of gender inequality is even more significant. In a majority of nations, 50% of the population is female; however, governments and policies do not represent this statistic. The rights of women are secondary to the economic advancement of men in both developing and developed nations. The disparity between genders increases as income decreases, meaning that women suffer from poverty more than men do. The discussion of financial access changes when distinguishing between genders. Achieving equal economic opportunity and access between genders poses the most monumental challenge to developing nations. However, the benefits of equality not only support the economic success of women but the success of the entire nation.

Inequality and Access in Developing Countries

Globally, 56% of women have financial access as opposed to 72% of men. The gap is smaller for developing nations at eight percentage points; however, only 67% of men and 59% of women have access (Demirgüç-Kunt, 2017). According to Demirgüç-Kunt (2017), women constitute about 56% of unbanked individuals. Unbanked individuals are typically less educated, lack documentation, and earn too little income to be able to open an account. Barriers to financial inclusion more adversely impact women than they do men, and in some developing nations, the gap in access reaches 30 percentage points (Demirgüç-Kunt, 2017). In some countries, sizeable gender gaps cause slow economic growth. The authors emphasise that any efforts to increase access must prioritise female inclusion (Demirgüç-Kunt, 2017).

In the 2020 WEF Gender Equality Report, economic participation is the second-largest gender gap at 57.8%, which has decreased from 2019. The report attributes the gap in economic participation in most countries to the lack of access to credit, collateral, and other financial services (WEF Global Gender Gap Report, 2020). Decreased access reduces the ability of women in developing nations to start or grow businesses and eliminates opportunities to save for the future. Given the rate of change over the past 15 years, the report estimates it will take 257 years to achieve gender parity. Next to political participation, economic empowerment is a verifiable method of encouraging equality as well as economic growth in underdeveloped economies. In the realm of accessing credit, 25 countries out of the 153 studied in the report do not have full inheritance rights and 72 countries do not have the right to open bank accounts or obtain credit (WEF Global Gender Gap Report, 2020). Women in both developed and developing nations conduct the majority of unpaid domestic work. Even in nations with the lowest disparity in unpaid housework, women still complete double the amount of work as men. It is evident from this report that a combination of government policy and social-cultural norms create inequality. However, increased financial access could be a critical method of reducing disparity.

van Staveren (2001) analyses gender bias in finance; noting that at the micro-level men and women are comparatively different on financial behaviour. Several factors contribute to the difference in behaviour, but van Staveren (2001) highlights the income gap as the most significant. In developing countries, the majority of women participate in unpaid work, such as household care and child-rearing, or find employment in low skill and low wage work. The initial factors of employment and pay give an unequal basis for financial inequality to grow further. van Staveren (2001) finds 'gender distortions' in financial markets, which create a disadvantage for females as they also lack collateral to participate in loan activity. Overall financial literacy analysis in developing nations; however, the gap between genders persists. Men are more financially literate than women are purely because they have greater access and therefore, experience in financial services. When women can actively invest and participate, they invest responsibly and have higher repayment rates on loans than men, assuming the woman retains control after receiving the loan. However, the author notes that in a study conducted on the Grameen Bank, only 37% of female borrowers retain control of their loans (van Staveren, 2001). The study of economic behaviour is necessary to construct policies that are inclusive and encourage financial access. Empirical evidence suggests that while women might not own collateral in the form of land titles, they are more likely to repay loans; thus, making them ideal candidates to increase financial services in developing nations.

Pallavi Chavan (2008) analyses the "extent and nature of gender inequality in the provision of banking services in India" (Chavan, 2008). The author uses two indicators to represent access to banking services: level of credit supplied to women and the level of deposits received by women. Chavan (2008) breaks down the data into subgroups to compare access between rural and urban areas; and finds a significant disparity between the two. Women in urban areas have significantly better access to financial services than women in rural areas have. Geographic location plays a vital role in a woman's ability to access financial services. Travel time, safety, and potential loss of productivity discourage women from accessing financial services, especially in rural areas. The study also found that proportionately, women contribute more deposits than they receive credit loans than men from financial institutions. Females can access financial services but are still limited in participation, highlighting the critical issue of inequality between genders. There are a few reasons men take out more loans than females: men are more willing to take risk, men make the decisions for the whole household, and men are more likely to take out loans to start businesses. Overall, Chavan (2008) points out key factors that contribute to unequal access to financial services between genders, but within females as well.

According to Seema Jayachandran (2015), unequal access to financial services and limited participation stem from more profound cultural practices that limit female autonomy to utilise services actively. The gap, caused by a lack of personal autonomy and participation in the workforce, is further exacerbated by non-existent financial independence. In Northern Africa, the Middle East, and India social-cultural norms restrict women's independence and physical mobility which in turn deepens unequal education, employment, and financial freedom. Traditional gender roles perpetuated by religious and cultural practices hinder the development of nations. It is essential to include the dimensions of inequality when analysing unequal access to financial services in developing nations. Country-specific characteristics create initial inequality that perpetuates unequal financial access. Gendered social constructs significantly influence the behaviours of women in developing nations. Physical mobility and independence, coupled with inequality in the eyes of the law, make accessing finance a dangerous act. Violence against women is common in developing nations and often occurs when women do not abide by social norms. If women do not feel safe accessing finance, then equal access will never occur. Some policies are attempting to circumnavigate the safety issue by providing mobile accounts allowing women to take control of their futures without waiting for society to change first.

A report by John Issac (2014) states, "Women disproportionately face barriers to accessing finance that prevents them from participating in the economy and improving their lives" (Issac, 2014). Female-owned businesses make up around 40% of small to medium firms in developing markets; however, a majority of the financial needs are unmet by current financial services. Due to the industry that most of these businesses operate, financial access is limited, but governmental and social institutions contribute to poor access. Female-owned organisations specialise in low skill sectors, and their ability to compete is limited. If given the same education and fair access, the most productive businesses and most innovative ideas will succeed. Policies must improve resources for women in business if economies seek to become competitive on the world stage and attain visible growth.

Klapper and Parker (2011) conduct a review on firm ownership and analyse the difference between male and female-owned firms. Male and female-owned firms differ between skill levels and industry types as well as gender. The majority of female entrepreneurs operate in labour-intensive, low skill, and highly competitive sectors. Businesses in this sector rely heavily on informal financial services and require less funding than a capital intensive high skilled operation, especially for developing countries. However, the direction of the relationship is undetermined; however, barriers in access could be driving women to pursue business ventures in low skilled industry leading to the distortion. According to the authors, there is no explicit discrimination from the financial institutions found, but institutional and social factors create most barriers to entry. Klapper and Parker (2011) note there is some evidence that women did face higher interest rates for the loans, but other factors like less start-up capital provided by the entrepreneur could account for the higher interest rate. When women do receive loans for business, there is a positive effect on economic growth in the nation. The authors' analysis emphasises the need to well-rounded quantitative analysis that includes country-specific development indicators.

Inequality and Access in Sub-Saharan Africa

Blackden et al. (2007) discover gender-based barriers that impede economic growth in Sub-Saharan Africa, where poverty is the highest among developing nations. Gender inequality spans multiple dimensions, but education, employment, control of assets, and governance issues are the significant areas highlighted by Blackden et al. (2007). All dimensions significantly contribute to gender inequality in SSA; however, the authors did not include financial access. Sub-Saharan Africa experiences such sparse growth due to poor infrastructure and policy. Simultaneous improvements to education, physical infrastructure, financial access, and governance are necessary to experience meaningful growth. The freedoms and rights of women vary by country and region, but the common thread that ties SSA together is the underutilisation of resources; including human, physical, and technological capital. The mutually beneficial relationship of financial access and human development indicators allows for a compounded effect that could drive significant growth in the region.

Ambreena Manji (2010) analyses development policy aimed at increasing participation in financial services for Eastern Africa and the effect the policies have on gender inequality in the region. The author notes that previous World Bank reports emphasise reforming the financial sector to promote growth and access to financial services. The use of land titles as collateral for accessing credit and participating in financial services is one main suggestion by the World Bank. However, inheritance laws and low wages exclude women from ownership, both legally and financially. Policies seeking to increase financial access by using tangible assets like land titles as collateral actively exclude women and the impoverished. Thus, this policy, and others like it, continue a pattern of institutional gender inequality in SSA and increase the adverse long-term effects of limited access. Manji (2010) argues that this biased policy perpetuates the cycle of inequality resulting from socio-cultural practices. The author raises concern over well-intended policies put forth by supranational agencies like the World Bank that lack the insight necessary to reduce negative spill overs. Current policies choose short-term improvements at the cost of long-run equality and inclusion.

Toyin Segun Ogunleye (2017) researches microfinance and the role of women in improving participation in financial services in Nigeria. The author suggests financial inclusion as a leading solution to decreasing poverty for men and women. In the analysis, women tend to be less risky and exercise more caution than men in financial decision-making. Given this assumed behaviour, Ogunleye (2017) test the instances of female repayment of microfinance loans. The results prove that increased microfinance loans to women in Nigeria improve the repayment rates and decrease the lender's risk. Increased microfinance for females not only reduces the gender inequality created by poor access but also encourages microfinance as an efficient method of reducing economic and gender inequalities.

Johnson and Nino-Zarazua (2011) investigate financial access and exclusion in Kenya and Uganda, emphasising country-specific research that encompasses social institutions and other unique variables. The authors find that factors such as employment, education, gender, and income heavily influence access to formal financial services. There is no difference in access between rural and urban residents. About 21% of adults in SSA have a mobile money account, which reduces the influence of geographic location to access. Johnson and Nino-Zarazua (2011) highlight the importance of using country-specific studies and condemn the one size fits all policy campaigns that are usually favoured by supranational organisations. The authors also investigate the level of formal financing by breaking down the dataset analysis to informal, semi-formal, and formal, which capture significant cultural and socioeconomic factors. Empirical findings support gender inequality favouring men in accessing formal financial services; however, in semi-formal and informal sectors, access is higher for women than men. These findings prove the importance of utilising country-specific analysis to capture certain cultural and social practices and improve accuracy. Gender inequality in financial access depends on the nation and its unique characteristics.

Aterido et al. (2013) investigate the differences in financial service accessed by households; breaking down their analysis between equality of financial services between personal accounts and business accounts. Access to finance for business includes formal and informal financing and the difference between genders in accessing formal financing channels. Initially, findings suggest that there is a gender gap, however, once controlled for external characteristics the authors find no difference between male and female access of formal financing for business (Aterido et al., 2013). Firm characteristics appear to create the inequalities present; especially size.

Selection bias contributes to observed inequality, as female entrepreneurs have higher barriers to entry than their male counterparts do. The authors find a definitive unconditional gap in access; however, there is no inequality between genders at the household level in use of formal services once controlled for other characteristics. Aterido et al. (2013) suggest that financial services themselves do not create inequality; instead, external characteristics influence the ability to access formal financial services. Education, income level, employment, and age are all key characteristics that affect accessing formal financial services.

Asiedu et al. (2013) note that current empirical findings suggest a lack of financial access is the primary constraint for all firm growth in Sub-Saharan Africa. Private enterprises remain underdeveloped due to unmet funding and economic instability. Female-owned firms face more constraints than male-owned firms do, and SSA firms are more constrained than any other region. Thus, female firms in SSA encounter the most adversity in financing. Asiedu et al. (2013) found that SSA female-owned firms experience 5.2% more in constraints than male-owned firms. After controlling for other characteristics and bias, the gender gap persisted, highlighting the severe disparity in financial access.

Hansen and Rand (2014) test the accuracy of survey data in determining unequal credit access of SSA firms by gender. Self-reported survey data measures the perceived inequality between male and female-owned businesses in accessing credit. The authors find a marginally significant gap between firms when using self-reported data (Hansen & Rand, 2014). However, when using formally reported data from financial institutions, no gender gap is found between male and female-owned firms. Hansen and Rand's (2014) research addresses discrepancies in measuring gender inequality in business, but access data is limited, and surveys are the best method of capturing actual and perceived access. Self-reported measures shed light on how people view access to finance and capture information on education, age, income, location, and other qualitative factors. The discrepancy between survey data and institutional data prove the influence of institutional factors on access and refocus the goal of increasing access back to reducing gender equality in all dimensions.

Nwosu and Orji (2017) analyse the impact of gender and formal credit access on small to medium-sized firm performance in Nigeria. Credit constraints reduce productivity and reinvestment into SMEs. The effect of credit constraints becomes more significant for female firms than male firms once adjusted for estimator biases. Formal access to credit has a significant impact on firm performance. Informal access to credit does provide a finance option that would otherwise not exist; it is not a reliable method of increasing firm growth and productivity. Equal access to formal financial services allows the most productive and competitive firms to grow. Nwosu and Orji (2017) suggest that policy reforms are necessary to increase access to formal credit loans and that will, in turn, improve performance of SMEs, especially female-owned firms in Nigeria.

Gender and Household Access to Financial Services in Rwanda

Mukamana et al. (2016) evaluated Duterimbere microfinance and its role in increase equal access to credit loans in Rwanda. In Rwanda, only one out of five women holds an account at a formal financial institution as opposed to four out of five men (Mukamana et al., 2016). The authors suggest the inequality in access arises from discriminatory policies against women. The main barrier to obtaining a credit loan requires collateral. The Rwandan government passed numerous initiatives, like Vision 2020, to increase access and reduce inequalities between genders. While tailored made programmes meant to increase equality, the benefit of the initiatives is unknown. Mukamana et al. (2016) conducted a series of questionnaires to determine loan amounts and frequency of accessing credit among women in Rwanda. The results reveal that microfinance initiatives improved equality in accessing loans through the tailoring requirements to increase the accessibility of credit. Eliminating collateral requirements and introducing group loans made access more widely available for women in Rwanda.

Gender Gap in Access to Finance in Rwanda

The importance of financial access in achieve gender parity is evident from the previously reviewed sources. Rwanda is no exception, and the case for equal access is bolstered by the nation's high global parity ranking and low economic opportunity rankings. Rwanda specific literature analysing gender equality and financial access are limited; thus, leaving much of the topic unexplored. Rwanda's gender parity performance is impressive, but financial access needs improvement for the nation to experience significant economic development.

According to the WEF Global Gender Gap Report (2020), Rwanda ranks first in Sub-Saharan Africa and ninth globally in gender parity. The nation experiences 79.1% gender parity but equality varies across measures of equality. Political empowerment in Rwanda ranks in the top four in the World, with over 50% of parliamentarians and ministers being female (WEF Global Gender Gap Report, 2020). Educational attainment is equal across primary and secondary enrolment; however, only 69.4% of women and 77.5% of men are literate. Overall tertiary enrolment analysis at 8%. However, men are twice as likely to pursue further education as women (WEF Global Gender Gap Report, 2020). The shortcoming in education translates to the lowest parity indicator, economic participation, and opportunity, which rests at 67.2%. Men and women participate equally in the labour force, but disparities arise in wages, professional employment, and leadership. A combination of unequal higher education enrolment and socio-cultural norms all contribute to economic inequalities. Rwanda is first in the World with perfect parity in labour force participation. The WEF Global Gender Report (2020) excludes financial access measures from parity calculations, potentially missing out on essential indicators of equality. The report suggests investments to education and human capital as primary methods to increase parity. While improvements to these areas are undeniable, financial access is proven to influence gender inequality; therefore, Rwanda's gender parity analysis is incomplete without it.

As noted by the WEF Global Gender Report (2020), Rwanda's educational enrolment is relatively equal across genders. Comparison of the primary and secondary gross percentage of enrolment rates across genders in Fig. 5.1 reveals that primary enrolment rates are relatively steady over time. However, there is a stark contrast between primary and secondary enrolment rates for both genders. While secondary enrolment increased since 2010, rates remain significantly lower than primary enrolment. The key take away from Fig. 5.1 is the difference in enrolment between genders. Both males and female access remain relatively equal over time, with most years experiencing higher female enrolment rates than male rates in both educational levels. Another critical aspect of gender equality in Rwanda is literacy. The 1994 genocide interrupted education form 1994 onwards until the late 1990s. The literacy rate captures adults over the age of 15 who can read and write; thus, including the children affected by the genocide.

Figure 5.2 compares the literacy rates of the adult population in Rwanda from 2010 through 2019. There is a consistent ten percentage point gap between male and female literacy rates with women experiencing lower

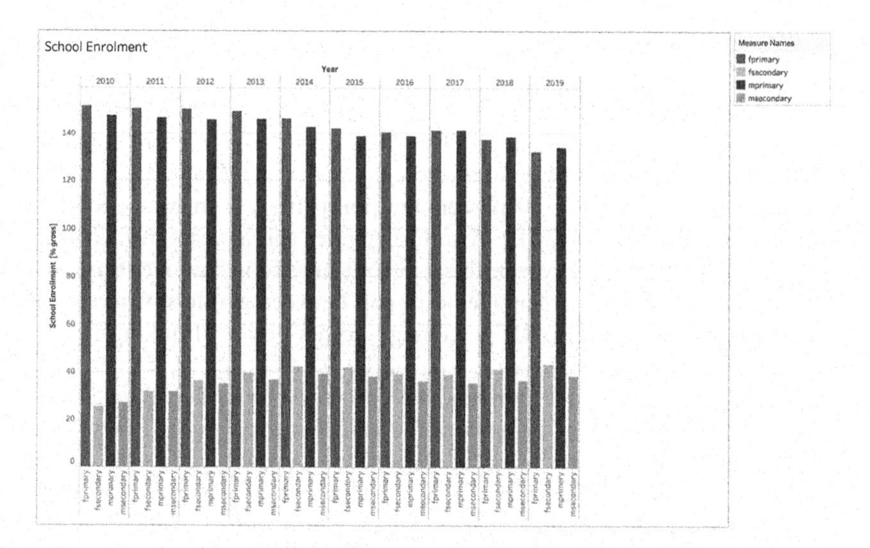

Fig. 5.1 School enrolment

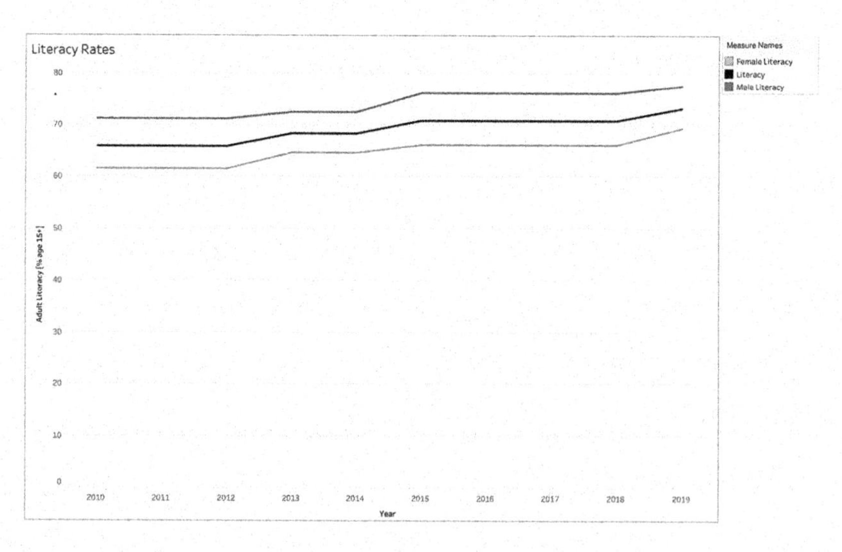

Fig. 5.2 Literacy rate

literacy. While the total literacy rate increased from 65% to 73%, women are consistently less literate than men are. Figure 5.2 suggests that gross enrolment rates are equal across the sexes; however, the difference in literacy casts doubt on this equality. The literacy rate may capture residual effects of gender inequality as well as the effects of the 1994 genocide.

Figure 5.3 illustrates a 20 percentage point increase in total account ownership from 2010 to 2019. A similar gender gap appears for account ownership in Rwanda; however, this gap continues to widen over time. In 2010, there is only a 5 percentage point gap between genders, whereas in 2015, the gap widened to 15 percentage points before shrinking to 10 percentage points in 2019. While total access is increasing over time, it is apparent that the increase is not equal across genders. For Rwanda to experience the long-term benefits of increased financial access, men and women should, at the very least experience equal growth rates.

Finally, Fig. 5.4 compares the type of accounts owned by males and females in Rwanda. A similar gap persists over time between the sexes for financial institution accounts. In 2015, total and male-owned financial institution accounts increased to 45% but decreased to 40% in 2019. Over the same period, mobile money accounts increase by 15 percentage points across both males and females. Increased access to mobile phones and the

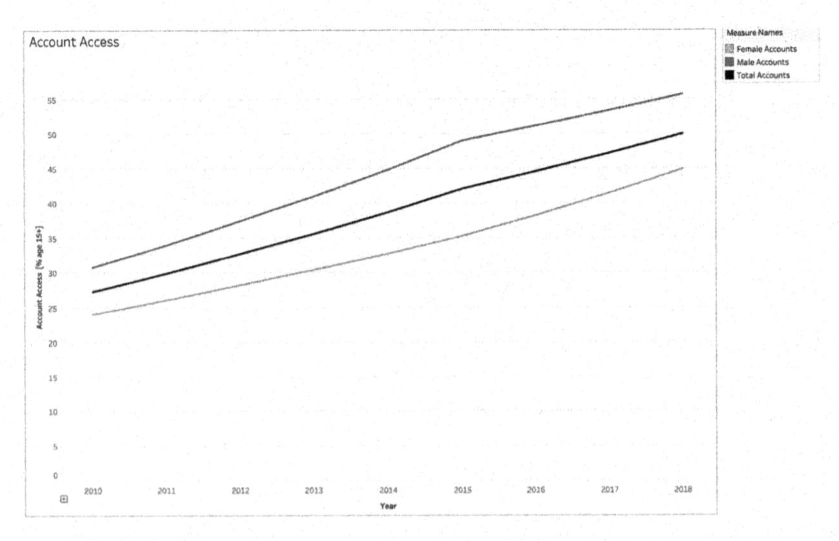

Fig. 5.3 Account access

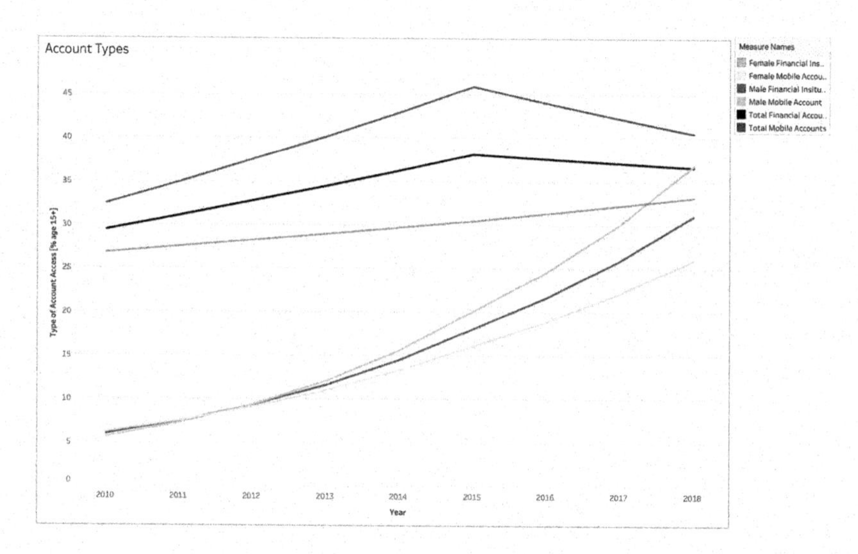

Fig. 5.4 Account types

internet also created a new method of account ownership in Rwanda. Should mobile money accounts continue this trend, mobile accounts would surpass financial institution accounts in the next few years. Mobile banking is a convenient method of storing, sending, and receiving money. Recent initiatives promote mobile money accounts as the most efficient method of increasing financial access in developing nations.

Gender gaps analysis of numerous dimensions; however, the gap in financial access is the most persistent. School enrolment rates suggest that younger generations might not experience the same gaps as older generations. However, disparities in literacy rates, wages, and employment skill levels suggest equality remains out of reach. Equal account access would ensure all people have a safe place to receive wages, save for the future, and actively participate in the financial economy.

Conclusion

Methods of development have evolved continuously over time and the factors that most affect development are finally coming to light. Development is a complex process that is unique to each country. The components that influence development range depending on the nation and there is no exact method that could spur growth for every country. Researchers have made an effort to determine some of the most important factors of development so that analysis for individual nations proves robust. Two main factors that seem to affect every developing and developed nation in the World are financial inclusion and gender equality. The unique social and cultural aspects of each country influence the variables, but their impact remains a standard for all. The goal of this research was to understand the relationships between gender equality, financial inclusion, and development of Rwanda.

Rwanda continues to make great strides in reducing gender inequality, and their high gender parity score is proof of their efforts. However, when analysing the differences in total account ownership, mobile account ownership, and financial institution account ownership, there is a noticeable gap between men and women. While education remains relatively equal, numerous dimensions like unpaid domestic work, labour skill level, and wages for similar work might lead to this inequality. However, until there is more data on these measures, it is impossible to determine the root cause of unequal financial access. For now, initiatives to increase gender equality and equal financial inclusion are appropriate methods to improve Rwanda's economic development.

References

Arestis, P., & Demetriades, P. (1997). Financial development and economic growth: Accessing the evidence. *The Economic Journal, 107*, 783–799. https://doi.org/10.1111/j.1468-0297.1997.tb00043.x

Asiedu, E., Kalonda-Kanyama, I., Ndikumana, L., & Nti-Addae, A. (2013). Access to credit by firms in sub-Saharan Africa: How relevant is gender? *American Economic Review, 103*, 293–297. https://doi.org/10.1257/aer.103.3.293

Aterido, R., Beck, T., & Iacovone, L. (2013). Access to finance in sub-Saharan Africa: Is there a gender gap? *World Development, 47*, 102–120. https://doi.org/10.1016/j.worlddev.2013.02.013

Blackden, M., Canagarajah, S., Klasen, S., & Lawson, D. (2007). Gender and growth in sub-Saharan Africa: Issues and evidence. In G. Mavrotas & A. Shorrocks (Eds.), *Advancing Development* (pp. 349–370). Palgrave Macmillan UK. https://doi.org/10.1057/9780230801462_19

Chavan, P. (2008). Gender inequality in banking services. *Economic and Political Weekly, 43*, 18–21.

Demirgüç-Kunt, A. (2017). The Global Findex Database 2017, 151.

Hansen, H., & Rand, J. (2014). Estimates of Gender Differences in Firm's Access to Credit in Sub-Saharan Africa. *Economics Letters, 123*, 374–377. https://doi.org/10.1016/j.econlet.2014.04.001

Issac, J. (2014). Expanding Women's Access to Financial Services [WWW Document]. World Bank. Accessed July 6, 2019, from http://projects-beta.worldbank.org/en/results/2013/04/01/banking-on-women-extending-womens-access-to-financial-services.

Jayachandran, S. (2015). The Roots of Gender Inequality in Developing Countries. *Annu. Rev. Econ., 7*, 63–88. https://doi.org/10.1146/annurev-economics-080614-115404

Johnson, S., & Nino-Zarazua, M. (2011). Financial Access and Exclusion in Kenya and Uganda. *Journal of Development Studies, 47*, 475–496. https://doi.org/10.1080/00220388.2010.492857

Klapper, L. F., & Parker, S. C. (2011). Gender and the Business Environment for New Firm Creation. *The World Bank Research Observer, 26*, 237–257. https://doi.org/10.1093/wbro/lkp032

Manji, A. (2010). Eliminating Poverty? 'Financial Inclusion', Access to Land, and Gender Equality in International Development: Eliminating Poverty? *The Modern Law Review, 73*, 985–1004. https://doi.org/10.1111/j.1468-2230.2010.00827.x

Mukamana, L., Sengendo, M., & Okiria, E. (2016). Promoting Gender Equality in Access to Microcredit Through Flexible Lending Approaches of Female Targeting MFIs: Evidence from Duterimbere MFI of Rwanda. *International Journal of Business and Economic Development, 4*, 12.

Nwosu, E. O., & Orji, A. (2017). Addressing Poverty and Gender Inequality through Access to Formal Credit and Enhanced Enterprise Performance in Nigeria: An Empirical Investigation: Addressing Poverty and Gender Inequality. *African Development Review, 29*, 56–72. https://doi.org/10.1111/1467-8268.12233

Ogunleye, T. S. (2017). Financial Inclusion and the Role of Women in Nigeria: Financial Inclusion. *African Development Review, 29*, 249–258. https://doi.org/10.1111/1467-8268.12254

van Staveren, I. (2001). Gender Biases in Finance. *Gender and Development, 9*, 9–17. https://doi.org/10.1080/13552070127734

World Economic Forum Global Gender Gap Report 2020 (Insight Report), n.d. World Economic Forum. http://www3.weforum.org/docs/WEF_GGGR_2020.pdf.

CHAPTER 6

Impact of Aid on Economic Growth in Nigeria

Douglas Onwumah and Bhabani Shankar Nayak

Introduction

Nigeria as a country is endowed with numerous natural resources like crude oil, coal, limestone, tin, iron ore, lead and zinc in addition to being blessed with a fertile land. Despite these natural resources, Nigeria has been identified as one of the poorest countries in the world, characterized by a high level of unemployment, low level of Income, high Inflation rate, low standard of living, hunger and malnutrition, etc. These economic challenges suggest the reasons why Nigeria is a major recipient of foreign aid.

Aid in Nigeria and other developing countries can be traced to the colonial era. Killick observed that a direct line can be traced from the

D. Onwumah (✉)
Adam Smith Business School, University of Glasgow, Scotland, UK

B. S. Nayak
University for the Creative Arts, Epsom, UK
e-mail: bhabani.nayak@uca.ac.uk

B. S. Nayak (ed.), *Political Economy of Gender and Development in Africa*, https://doi.org/10.1007/978-3-031-18829-9_6

creation of the Overseas Development Ministry (ODM) by the British government in 1964 for grants in aid provided to the colonies from the 1870s which was more formally organized under the 1929 Colonial Development Act and the Colonial Development and Welfare Acts of 1940 and 1945.

Fatukasi and Kudaisi observed however that the record of Western aid to Africa has been significant, amounting to more than $500 billion between 1960 and 1997 which is the equivalent of four Marshall Plans being pumped into sub-Saharan Africa. According to Fatukasi and Kudaisi (2015), in times past, the national budgets, apart from the relief aid and economic development, foreign aid assistance were also provided to support reforms and policy adjustment programs in some developing countries including Nigeria. For example, between 1981 and 1991, the World Bank provided $20 billion toward Africa's structural adjustment program and Nigeria was a beneficiary in 1986. In addition, Ogbe noted that Nigeria also benefited from debt cancellation, worth $107.2 million, by Canada and the United States in 1989/90 in addition to the $18 billion debt cancellation by the Paris Club in 2005.

Furthermore, Nigeria's foreign aid flow has increased over time from $39 million in 1981, $258 million in 1991 and $3.3 billion in 2017. Despite these increases, the economy is still characterized by low level of income, high level of unemployment, very low industrial capacity utilization and high poverty level (Fasanya & Onakoya, 2012). It is against this backdrop that I aim to determine the impact of foreign aid on Nigeria's economic growth. Although there are previous studies relating to aid and economic growth in Nigeria, none of these studies captures the sample period of 1981–2017, this may be attributed to non-availability of data.

A lot of studies have been carried out on the relationship that exists between foreign aid and economic growth. For Instance, Nwosu (2018) employed time series data from 1981–2016 to ascertain the effect of foreign aid on economic growth in Nigeria. She used two-stage least squared (2SLS) in her econometric analysis and found out a positive and significant, albeit marginal aid-growth relationship. The study further suggests that although foreign aid is important for economic growth in Nigeria, it is not among the economy's major growth drivers. To improve the effectiveness of foreign aid received by Nigeria, the author recommends the implementation of sound macroeconomic policies and the need for institution strengthening.

In addition, Mbah and Amassoma (2014) used time series data spanning from 1981 through 2012 to investigate the effect of foreign aid on the economic growth of Nigeria. The authors employed econometric techniques such as ordinary least square (OLS), Augmented Dickey-Fuller (ADF) test and Johansen cointegration test; they found a negative and nonsignificant relationship between foreign aid and economic growth. They attributed their findings to corruption and weak institution prevalent in the Nigerian economy; they further proposed the implementation of political, economic and institutional reforms that will address the problem of pervasive corruption in the country, improve the quality of governance, ensure that foreign aid flows are invested into developmental projects that will boost the nation's GDP and reduce the level of poverty in the country.

Okoro et al. (2019) examined the effect of international capital inflows on the economic growth of Nigeria with a data set spanning from 1986 to 2016. The study employed four core channels of international capital inflows which include foreign direct investment (FDI), official development assistance (ODA), personal remittances (REM) and external debt stock (EXTDS) into Nigeria as the explanatory variables and GDP growth rate as the dependent variable. They employed Johansen cointegration and ordinary least square (OLS) techniques for data analyses and discovered that international capital inflows have a long-run effect on the economic growth of Nigeria. Specifically, the OLS revealed that FDI and REM had significant positive effects on economic growth while EXTDS and ODA had no significant effects on economic growth. The study further recommends that policymakers should discourage the use of external debt and official development assistance in Nigeria.

Similarly, Kolawole (2013) examined the impact of foreign assistance (ODA) and foreign direct investment (FDI) on real growth in Nigeria from 1980–2011, using the two-gap model framework that is savings gap and foreign exchange gap and various econometric techniques which include Augmented Dickey-Fuller (ADF) test, Granger causality test, Johansen cointegration test and error correction method (ECM). The findings revealed that imports impact negatively on real growth, while domestic investments and exports had a positive effect on real growth in Nigeria. The main results from this study posit that while FDI showed a negative effect, ODA revealed an insignificant effect on Nigeria's GDP. The results establish that the bulk of foreign assistance meant for infrastructural development in the country is either siphoned or diverted into unproductive use, which indicates why the impact of aid on economic growth is hardly noticed.

More so, Fasanya and Onakoya (2012) employed a time series approach to examine the impact of foreign aid on economic growth from 1970–2010 using the neoclassical modeling framework. The authors employed the ordinary least square and error correction model estimation techniques to estimate the causal relationship between foreign aid and economic growth in Nigeria. Their findings show that while population growth has no significant effect on aid flows, aid flows yield a significant impact on economic growth in Nigeria and that there was an increase in domestic investment resulting from aid flows.

Burnside and Dollar's (2000) work is the most cited literatures on aid-growth nexus, which used a panel of 56 countries and six 4-year time periods from 1970–1993 to examine the relationship between foreign aid, economic policies, and the growth of per capita GDP. The study uses ordinary least square and two-stage least square estimation techniques and finds that aid alone if included in the model has no significant effect on real per capita GDP growth. They further noted that aid interacting with policy gives a positive and significant effect on growth, implying that aid will only improve growth if the recipient country has good fiscal, monetary and trade policies.

Siddique and Kiani (2017) studied the impact of foreign aid on economic growth using evidence from a panel of South and East Asian countries for the period of 1995–2013 using the dynamic panel estimation technique. The study showed significant and robust results that foreign aid promotes the economic growth for the countries in the panel and supports several empirical results on the positive and significant impact of external finance on domestic savings, investment, and growth, such as the study of Burnside and Dollar. They further conclude that foreign aid has a positive impact on economic growth in the underdeveloped economies if they have a good monetary policy, fiscal policy and trade policy, but in the absence of these policies, there is a small impact of aid on growth.

Mustafa et al. (2018) examined the relationship between foreign aid and economic growth in Sudan using autoregressive distrusted lag (ARDL) bounds tests for cointegration and a time series data spanning over the period of 1980 to 2015, their findings show that there is a long-run relationship between variables under consideration. Specifically, the findings show that foreign aid in the form of official development assistance (ODA) has a positive and statistically significant long-run impact on economic growth in Sudan, while the interaction between aid and corruption in public institutions is found to have a negative and significant long-run impact on economic growth, indicating the harmful impact of corruption in reducing the feasible contribution of aid to economic growth. The findings also indicate that aid deters economic growth in the short run.

Furthermore, Yiew and Lau (2018) investigated the role and the impact of foreign aid (ODA) on economic growth (GDP) using 95 developing countries as the sample. They included foreign direct investment (FDI) and population (POP) as the control variables, their results indicate that a U-shape relationship exists between foreign aid and economic growth. Initially, foreign aid negatively impacts the countries' growth and over a period it positively contributes to economic growth. Furthermore, the results strongly support the view that both FDI and POP are more important determinants of GDP, implying that GDP is less likely to depend on ODA. They find out that strengthening the legal framework of the sample countries would be essential while their overdependency on the influx of ODA might lead to negative impacts on the growth.

Tang and Bundhoo (2017) use panel data for ten SSA countries from 1990–2012 to investigate the impact of foreign aid on economic growth in ten sub-Saharan African countries with a data spanning from 1990 to 2012. They employ OLS, pooled OLS, fixed effects, random effects, first-difference estimator and 2SLS estimation techniques for robustness purposes. Their findings reveal that there is a nonsignificant relationship between aid and economic growth, however, aid interacted with policy showed a positively significant relationship, which suggests that the effectiveness of aid depends on good policies, these findings agree with Burnside and Dollar (2000). They also find that institutional quality has a significant positive effect on growth. Additionally, the authors test the two-gap growth model and conclude that foreign aid is a good supplement for investment and import requirement in the investigated SSA regions. The study ascribes the insignificance of aid to the unstable economic, institutional and political environment in these regions.

Liew et al. (2012) employ panel data methods namely pooled OLS, random effects and fixed effects to examine the impact of foreign aid on the economic growth of East African countries over the period of 1985 to 2010. The results of the finding reveal that foreign aid has a significant negative influence on economic growth for these countries. Ndi has argued that African countries depending on foreign aid as an alternative way to funding development through the creation of a new capital bond market, microfinancing, revised property laws and enhance political stability with the objective of attracting foreign commercial investments.

After reviewing some literature, it seems that there exists an inconclusive literature on the relationship between foreign aid and economic growth. For instance, Nwosu (2018), Mustafa et al. (2018), Yiew and Lau (2018), Siddique and Kiani (2017), and Fasanya and Onakoya (2012), report a significant positive correlation between aid and growth. Okoro et al. (2019), Tang and Bundhoo (2017), Liew et al. (2012), Mbah and Amassoma (2014), and Kolawole (2013) find evidence for a negative impact and insignificant effect of foreign aid on growth. The reason for these inconsistencies is still unknown in the body of research, though some attribute the disparities to inaccuracies in data, and differences in methodology. This study is aimed at examining if the relationship between foreign aid and economic growth in Nigeria exhibits a positive, negative or insignificant relationship.

Most literature on aid has been criticized based on the existence of philosophical gap and empirical gap. Burnside (1998) and Dollar (2000) have criticized the choice of policy variables and policy measure that makes it difficult to identify the relationship between policy and growth. In addition, most empirical results are not robust due to methodological weaknesses.

Aid in Nigeria

Foreign aid is the international transfer of public funds in the form of loans or grants either directly from one government to another (bilateral assistance) or indirectly through the vehicle of a multilateral assistance agency such as the World Bank (Todaro and Smith, 747–891). Olagboyega while explaining foreign aid noted that it is used to cover all financial transactions made or guaranteed by one government to another and has become a focus and locus in the Third World which is now used as a foreign policy instrument by developed democracies to strengthen their relationship with the developing countries and consequently spread their influence on them. Also, the establishment of aid was one of the principles of the Breton Wood system in 1914. The system believes that there should be a free capital market that allows for an unrestricted inflow of foreign aid. Based on this thinking, a Marshal Aid Assistance of about $17.5 billion was granted to European countries to resuscitate their ruined economies due to the Second World War. Since then, foreign aid has remained a veritable economic phenomenon of the international economic system (Ukpong, 2017).

Developmental aid has been the highest source of external funding in less developing countries including Nigeria. One major characteristic of these economies is the issue of budget constraint which has to slow down growth and development over time. The availability of development aids is believed to either relax budget constraint of a country or influence its expenditure. It can also work in stimulating economic growth thereby supplementing available sources of finance such as revenue, capital investment and capital stock of a country (Olanrele & Ibrahim, 2015). Also, Nigeria's over-reliance on petroleum has made the economy vulnerable. The prices of crude oil in the global market consequently leading to the recent decline in annual GDP growth rate from 3.52% in 2014 to -0.01% in 2015 which depicts the level of the country's vulnerability and has made the country an attractive destination for foreign aid.

According to the OECD's Aid at a Glance data, total net official development assistance (ODA) provided to Nigeria by the members of the OECD's Development Assistance Committee (OECD-DAC) amounted to USD 11.4 billion in 2006 (of which USD 1 billion was for debt relief) while the annual inflow of aid increased from $39 million in 1981, $258 million in 1991 and $3.3 billion in 2017. The OECD's data for 2005–06 show that the top four donors were the United Kingdom, France, Germany and Japan.

Conventionally, foreign aid is expected to add to the stock of physical, human and institutional capital of the recipient country, which might boost the productive capacity and growth of the economy. However, it has been argued that the failure of foreign aid to produce the desired result in the recipient countries is based on the recipient countries' absorptive capacity, corruption level and sound economic management.

In Nigeria and other LDCs, the effectiveness of aid has generated many debates in the aid-growth literature, for instance, Nwosu (2018), Mustafa et al. (2018), Yiew and Lau (2018), Siddique and Kiani (2017), and Fasanya and Onakoya (2012), found a significant positive correlation between aid and growth. While Okoro et al. (2019), Tang and Bundhoo (2017), Liew et al. (2012), Mbah and Amassoma (2014), and Kolawole (2013), found evidence for a negative impact and insignificant effect of foreign aid on growth. The reason for these inconsistencies is still unknown in the body of research, though some attribute the disparities to inaccuracies in data and differences in methodology.

History of Aid in Nigeria

The transatlantic slave trade in the fifteenth century which was pioneered by the Portuguese paved the way for colonization in Nigeria and other African countries. The Portuguese activities continued until the British replaced the Portuguese as the leader of the slave trade business in the eighteenth century. The British traders had settled in Nigeria around this time in an area that surrounded the Niger River known as Lagos, the subsequent abolition of the slave trade was the key moment when the British truly "intervened in the region." They placed their focus on obtaining goods to increase their ability to trade. Lagos was a colony in 1861 which later paved the way for the amalgamation of the Southern and Northern protectorate in 1914.

Since that union, different forms of aid have been provided by the British government. In addition, different laws were promulgated toward the proper administration of aid in their colonies. For example, under the Colonial Development Act of 1929 and Welfare Acts of 1940 and 1945, the British government through the Overseas Development Ministry administered grants for infrastructural development and the expansion of social sector activities across her colonies, which objective was centered on gaining political control over the colonial domain. Brett noted that aid was used by the colonial powers to build the needed infrastructure to sustain local raw material exports and their own manufactured exports in return and relied heavily on missionaries to run schools and health systems.

Foreign aid and other forms of development assistance have been a permanent feature of the political economy of post-colonial African states like Nigeria; this was because of the developmental challenges faced by Nigeria after her independence on October 1, 1960. Immediately after independence, the development paradigm was centered on government controls and national planning, which include building the basic infrastructures on which the economy can thrive, this is in addition to building the basic hard and soft infrastructure necessary to restructure the economies and to create the right environment for her citizens to enjoy the benefit of freedom. More so, investment in key physical infrastructures such as bridges, electric power plants, dams, networks of roads, railway systems, and the establishment of industries modeled after the developed countries became the main concern of some African countries. During this period, Nigeria like other African countries did not have enough financial resources to carry out these key national development projects. The international community role was to support the newly independent states with grants, loans and other forms of technical assistance (Kalu, 2018).

In a view of speeding up development, various policies and plans were initiated in Nigeria. These plans include The First National Development Plan (1962), The Second National Development Plan (1970–74), The Third National Development Plan (1975–80) and The Fourth National Development Plan 1981–85. Along these lines, most foreign aid that Nigeria received in these periods was geared toward supporting the implementation of the national development plans. For instance, two years after independence, the first National Development Plan policy was formulated between 1962 and 1968 with the goal of development opportunities in health, education and employment and improving access to these opportunities, etc. external financing from loan, aid and grants was expected in order to finance the plan (Lawal & Oluwatoyin, 2011). In addition, in these development plan eras, the World Bank and order notable multilateral financial institutions were an important channel of support. Bilateral supports also came from advanced nations. The principal donors include the World Bank, the United States Agency for International Development (USAID), the U.K. Department for International Development (DFID), the African Development Bank (AfDB), the European Union, the Canadian Development Agencies (CIDA), Japanese International Co-operation Agency (JICA), the French Agency for International Co-operation, and the United Nations. The first five donors account for over 80% of annual development aid to Nigeria (WTO, 2008).

Kalu (2018) observed that in the early 1990s, Nigeria and other African countries witnessed direct interventions by donors using international and local NGOs, which has worked most importantly in the area of public health, where donors have directly provided services designed to eliminate malaria, provide vaccinations, or to support the prevention and treatment of HIV/AIDS. He further noted that the rationale for direct interventions by donors was the feeling that governments sometimes divert aid funds to other activities that are not welfare-enhancing which made donors feel that by delivering services directly through the use of NGOs, the risk of mismanagement of donor funds would be reduced, and in the process, foreign aid would be more efficient and effective in achieving its aims.

Theories of Aid in Nigeria

According to Pankaj AK in his article "Revisiting Foreign Aid Theories" published in 2005, "Development economics does not recognize the independent existence of foreign aid theories. They are taken to be a part of the general theory of growth and development and as most of the aid theories, which are employed today, are variants of, or a logical development of the various theories of growth and development they are not considered independently." In support of this proposition, this study will look at three theories of aid with a view of understanding the theoretical and philosophical foundation of aid in Nigeria. These theories include the big push theory, the dependency theory and the theory of the white man's burden.

The Big Push Theory

The big push theory is one of the earliest theories of development economics which was propounded in the 1940s by Rosenstein-Rodan P.N. According to him "There is a minimum level of resources that must be devoted to … a development program if it is to have any chance for success. Launching a country into self-sustaining growth is a little like getting an airplane off the ground. There is a critical ground speed which must be passed before the craft can become airborne."

The principle behind the theory holds that developing countries need a high minimum level of investment to kick start economic development and until such investment is made achieving higher levels of productivity and income cannot be guaranteed. The crux of the theory is that only a big and wide-ranging investment package engineers and stimulates economic development which highlights that a definite amount of resources should be dedicated for developmental programs (Umoru & Onimawo, 2018). Accordingly, the "BP" theory further stresses that a "bit by bit" investment program will not impact the growth process required for development but rather a large amount of investments to propel the path of economic progress from a contemporary state of backwardness. Hence, the theory became the justification for foreign aid. Kilman and Lundin (2014) observed that many economists advocated a big push involving a combination of a large increase in aid and a simultaneous increase in investment in numerous sectors leading to economic growth and poverty reduction. Foreign aid has different purposes and is operated through different channels, humanitarian and food aid for instance goes directly to household units, the impact of which is felt on a micro level while development aid mainly finances government budgets, and public investments are meant to have a macro impact on the economy.

In support of the big push theory in Nigeria, the pre-colonial and colonial aid system in Nigeria was aimed at building infrastructure and social services; this is in addition to the various bilateral and multilateral aids and grants Nigeria has benefited from since independence. The policy implication of the big push theory in Nigeria is anchored on the need for sound policy which has affected the effectiveness of aid in Nigeria. Umoru and Onimawo (2018) observed that foreign aid to Nigeria has not been effectively managed to promote investment and growth in the economy which is attributed to be due to corruption and aid fungibility. The authors further conclude that sound macroeconomic policies should be put in place to ensure that foreign aid flows are invested into developmental projects that will boost the nation's GDP and reduce the level of poverty in the country.

In a research by Burnside and Dollar in 2000 they emphasized that growth of developing countries to a large extent depends on their economic policies and that foreign aid has not really achieved its objective of expanding growth. The empirical research by Collier and Dehn, Collier and Hoeffler and Collier and Dollar have identified the effectiveness of aid in a good policy environment.

The Dependency Theory

The dependency theory was developed by Raul Prebisch in the late 1950s; Prebisch and his colleagues were troubled by the fact that continuous and sustained economic growth experienced by the rich and developed countries failed to influence growth in the poorer countries. They are of the view that the less developed countries were exploited because of global capitalism, which makes them dependent on rich countries (Ferraro, 2008). The developing countries exported primary products to the rich countries which have industries and technological know-how, the developed countries in return process and manufacture finished products out of those primary products and then sold them back to the poorer countries thereby making the developing countries stagnant at the expense of the rich industrialized countries. He further noted that the main ideas of the dependency theory are that developed countries benefit heavily from the resources of poorer nations which enables the richer countries to sustain higher standards of living. In addition, Okpanachi (2011) noted that dependency theory explains the factors responsible for the position of the Third World countries and their constant demand for, and continuous reliance on aid from the developed countries.

Frank explained that dependency theory emerged because the orthodox development theories like the modernity theory fail to capture the nature of relationship that existed between the rich nations and the poorer regions of the world which he argues was rhetoric as they do not unveil the objective behind the giving of donor aid to undeveloped countries. Kabonga observed that dependency theory is premised on resource mobilization, which involves the flow and movement of resources from poorer countries, referred to as the periphery, to rich countries that are referred to as the core—this flow of resources enriches the core while impoverishing the periphery, these core states he further noted are rich and industrialized states in the Organization of Economic Cooperation and Development (OECD) while the periphery states are countries in Latin America, Africa and Asia which are characterized by low per capita gross national products (GNPs) and are reliant on a single commodity to earn foreign exchange like the case of Nigeria which solely depends on oil as her foreign earning.

In the Nigeria colonial period, the British dominated and controlled the Nigerian economy, the system favored production of cash crops and other raw materials such as tin and gold for British industries and in exchange Nigeria received imported manufactured goods. The colonial protectionist policy operated during the colonial era ensured that exports were restricted to the Britain only and through the aid of colonial merchants' companies such as the United African Company (UAC), the United Trading Company (UTC), and the African Timber and Plywood Company (ATP); these companies acted as cartel thereby monopolizing trading at the expense of the local merchants.

Nnoli in his analysis of British colonialism and its overall impact on Nigeria's underdevelopment asserts thus:

...the policy of the integration of pre-colonial Nigeria into the global capitalist economic system, as a peripheral member by the colonialists, caused the destruction of the society's rich and varied political systems, and social structure, and the creation of new productive economic activities based on the need of foreign capitalist countries. It diverted attention away from local creative potential and resources by focusing on the production of primary resources needed by Europeans.

The Theory of The White Man's Burden

Willian Easterly's book "*The White Man's Burden*" was a repeat of his argument in his first book titled "*The Elusive Quest for Growth.*" He examines the objectives and effectiveness of past efforts by the rich nations to reduce suffering and thus eliminating the extreme poverty being experienced by the poor countries. According to Easterly, foreign aid doesn't work because it fails to consider the fact that people respond to incentives. For example, if you spend lots of money on education, but there are no jobs, people will have little incentive to remain in school so enrolment will remain low (McMillian, 2007). In reviewing the effectiveness of aid to the developing countries, Mccloug noted that despite five decades of rhetoric and intricate plans by intelligent people with noble intentions, and despite $2.3 trillion in donor aid, extreme poverty persists in the developing world. The theory according to Easterly is that the aid world is divided into two groups: planners and searchers—while planners create grand top-down schemes to address economic issues, searchers on the other hand focus on specific projects that are locally based and incorporate community involvement that is result oriented. The idea of the theory is for donor's fund to be judiciously accounted for.

In support of the theory of the white man's burden, Bain et al. (2016) observed that there have been periodic calls for donors to change how aid is provided since the 1960s, with a common emphasis on giving more decision-making power to beneficiaries and providing aid in ways that take better account of the local context. Furthermore, at the beginning of the twenty-first century, it became clear that increases in aid financing were not producing the impact expected. While interest in aid effectiveness was not new, an unprecedented consensus emerged on what needed to be done to produce better results leading to the Paris conference of 2015 and now the Paris declaration on aid effectiveness (PD) (KPMP Report, 2016). In ascertaining the effectiveness of foreign aid in Nigeria, some studies have concluded that aid has not really impacted the economic growth of Nigeria judging from the huge aid inflow. For instance, Olanrele and Ibrahim (2015) concluded that with a view to promoting the meaningful impact of developmental aid on economic growth in Nigeria there is need by the donors to provide standard monitoring and evaluation framework.

Chukwuemeka et al. (2014) acknowledged that many factors were found to militate against the effectiveness of foreign aid in achieving development, according to them, these factors are, including corruption, poor policies and institutional framework as well as poor utilization of development fund.

Trends of Aid in Nigeria

Many developing countries are faced with a lot of economic challenges which have slowed down growth and development, some of these challenges include budget constraint, recurrent expenditure highly exceeding capital expenditure, two-digit inflation rate, high unemployment rate, unfavorable balance of payment and many more necessitating the importance of adopting foreign assistance as a source of financing. According to Olanrele and Ibrahim (2015), the availability of development aids is believed to either relax the budget constraint of a country or influence its expenditure, which in addition can also work in stimulating economic growth, thereby supplementing available sources of finance such as revenue, capital investment and capital stock of a country.

From 1981 to 2017 Nigeria has witnessed ten changes of government, including both military and democratically elected governments. This includes President Shehu Shagari (1979–1983), Major-General Muhammadu Buhari (1983–1985), General Ibrahim Babangida (1985–1993), President Ernest Shonekan (1993), General Sani Abacha (1993–1998), General Abdulsalami Abubakar (1998–1999), President Olusegun Obasanjo (1999–2007), President Umaru Musa Yar'Adua (2007–2010), President Goodluck Jonathan (2010–2015), President Muhammadu Buhari (2015–till date). Within these periods, there has been a massive inflow of foreign aid from both multilateral and bilateral sources. For instance, during General Ibrahim Babangida's era as the military head of state from 1985 to 1993, in his policy framework abandoned development planning and opted for greater neo-liberal market approach and private sector-driven development strategy known as the Structural Adjustment Program (SAP) which was an International Monetary Fund (IMF) and World Bank-supported short-term package that was expected to last till June 1988 but stretched to 1993 (Abah & Naakiel, 2016). Within this period, foreign assistance amounted to about US$17 billion.

Furthermore, during General Olusegun Obasanjo's regime between 1999 and 2007 there was a drastic increase of foreign aid, especially from

2005 to 2007, when Nigeria benefited from a debt relief worth $18 billion from the Paris club and an overall reduction of her debt stock by $30 billion, thereby positioning Nigeria to become the second largest recipient of Official Development Assistance and was ranked among the tenth highest recipient in Africa. According to the World Bank, the average volume of Net Official Development Assistance and official aid received by Nigeria between 1981 and 1990, 1991 and 2000, 2001 and 2010, 2011 and 2017 ranges from US$102 million, US$212 million, US$2.6 billion and US$ 2.4 billion respectively while the Net Official Development Assistance received (% of GNI) witnessed an average of 0.63% from 1981–2017. However, this inflow has not been judiciously utilized. Olanrele and Ibrahim (2015) observed that foreign aid looted in Nigeria is equivalent to total aid to other countries in Africa over the past decades.

Evidence from Studies in Nigeria

For capital deficient countries like Nigeria as observed by Delessa (2012) in the case of Ethiopia, where there is a low level of domestic saving which makes it difficult for the economy to meet the requirements of demand for investment, the significant role foreign aid plays in financing the resource gap is undeniable as noted by Chenery and Strout (1966) in their two-gap model. More so, in a way to ascertain the impacts of foreign aid in Nigeria, different studies have been undertaken with different results, policy viewpoints and recommendations.

The most recent study by Orji et al. examines the contributions and impact of foreign aid on capital formation in Nigeria by employing the autoregressive distributed lag (ARDL) model and Granger causality test. The study shows that foreign aid (proxied by ODA) has a positive and significant impact on capital formation in Nigeria while the result of the Granger causality test shows that a bi-directional Granger causality exists between foreign aid and gross fixed capital formation (GFCF). They further recommended that the government should expedite actions toward the implementation and effective utilization of foreign aid, this is in addition to the development of good policy measures that would monitor the maximum and effective utilization of foreign aid.

More so, Okoro et al. (2019) examined the effect of international capital inflows on the economic growth of Nigeria with a data set spanning from 1986 to 2016. The study employed four core channels of international capital inflows which include foreign direct investment (FDI), official development assistance (ODA), personal remittances (REM) and external debt stock (EXTDS) into Nigeria as the explanatory variables and GDP growth rate as the dependent variable. In their analysis, they employed Johansen cointegration and ordinary least square (OLS) techniques for data analyses and discovered that international capital inflows have a long-run effect on the economic growth of Nigeria. Specifically, the OLS revealed that FDI and REM had significant positive effects on economic growth while EXTDS and ODA had no significant effects on economic growth. The study further recommends that policymakers should discourage the use of external debt and official development assistance in Nigeria.

Furthermore, Fashina et al. (2018) applied a two-model approach to investigate the link between aid and human capital in promoting the economic growth of Nigeria, the first model was used to test the validity of the medicine model in Nigeria, while the second model was used to investigate the effect of aid and human capital shocks on growth using Engle-Granger and vector error correction model (VECM) estimation techniques. Their findings suggest that a persistent increase in foreign aid flows beyond a particular time (the optimal point) may adversely affect growth thus confirming the proposition of the Medicine Model. In addition, evidence from the second model indicates that growth in Nigeria is sensitive to human capital shock via education while the response from aid shock is trivial in the long run. The study observed that although the attainment of economic growth might be challenging for an aid-dependent country like Nigeria, it further concludes that government expenditures on education with additional inflows of aid can promote economic growth in Nigeria.

In addition, Babalola et al. observed that the inflows of foreign direct investment, foreign aid and foreign trade have been on the increase in Nigeria in the past three decades while the relationship between these variables and economic growth has not been thoroughly explored. Hence, they examined the impact of foreign direct investment, foreign aid and foreign trade on economic growth in Nigeria using annual time series data covering the 1980–2015 period and by employing the autoregressive distributed lag (ARDL) model-bounds test as well as the error correction model (ECM); they found out that the variables are cointegrated, they further concluded that foreign direct investment, foreign aid and foreign trade have positive long-run impacts on economic growth in Nigeria while in the short run, only foreign aid has a positive impact on economic growth. In addition, they suggested that to accelerate economic growth, the government should strengthen policies that can accelerate foreign direct investment, foreign aid and foreign trade, and further opening up of the economy and the promotion of greater cooperation with development partners will enhance economic growth in Nigeria.

In examining the effect of external borrowing and foreign financial aid (ODA) on Nigeria's economic growth for a period of 1980–2013, Ugwuebe et al. employed ordinary least square technique (OLS), Augmented Dicker-Fuller (ADF) test, Johansen cointegration test and error correction method (ECM) as the econometric techniques, the authors find evidence that in the short run, exchange rate and foreign reserve exhibit positive and significant impact on economic growth in Nigeria, while total grants (ODA) revealed a positive and significant relationship. However, in the long run, external debt was positive and significant while ODA revealed a positive but insignificant result. The authors explain that aid had a positive impact on economic growth in Nigeria but is insignificant because foreign aid is expended on recurrent expenditure rather than productive investment or capital expenditure.

Olanrele and Ibrahim (2015) examine the effect of four different types of developmental aid (multilateral aid, bilateral aid from Nigerian's trading partners, bilateral aid from the top-five CDI ranked countries and bilateral aid from Nordic countries) on economic growth in Nigeria using a time series data spanning from 1970 to 2012. The study shows that multilateral aid had more impact on growth compared to bilateral aid from Nigerian's trading partners, top-five CDI ranked countries and Nordic countries. The study further concludes that with a view to promoting the meaningful impact of developmental aid on economic growth there is a need by the donors to provide standard monitoring and evaluation framework.

Bakare et al. (2014) looks at the macroeconomic impact of foreign aid in sub-Saharan African (SSA) countries using Nigeria as a case study. The author uses the vector autoregressive (VAR) model to determine the sources of a shock to growth in Nigeria and treats foreign aid as an endogenous variable. The findings of this study reveal a negative relationship between aid and output growth, implying that foreign aid worsens Nigeria's output growth. The author also believes that the study confirms with his earlier study on aid fungibility where he argues that foreign aid did not promote growth in sub-Saharan Africa.

Many studies with divergent views have been carried out on the impact of foreign aid on economic growth, below are the summaries of empirical literature reviewed on the relationship between foreign aid and economic growth outside Nigeria.

Hussain et al. investigated the impact of foreign aid inflows on economic growth in Pakistan, India, Bangladesh and Sri Lanka of SAARC with a specific focus on foreign aid and economic growth nexus in Pakistan with other countries. The study found that foreign aid inflows put negative effects on the economic growth of Pakistan and all other SAARC countries. Education, gross capital formation and population growth rate exert a positive relationship with economic growth, whereas as expected inflation was insignificant. Furthermore, the findings suggest that Pakistan and other SAARC countries should efficiently utilize domestic resources instead of dependence on foreign aid for accelerating economic growth.

Sothan examined the growth impact of foreign aid in Cambodia over the period 1980–2014 using the autoregressive distributive lag (ARDL) model while incorporating investment and trade openness as part of the model. The study shows that trade openness has positive effects on growth in both the short run and the long run while investment has positively contributed to growth in the long run. In addition, foreign aid has a positive impact on growth only in the short run while in the long run, it has a negative impact on investment and growth. The study further suggests that policymakers should promote investments through elevating domestic and foreign capital in the country rather than total dependence on aid.

In addition, Walin (2014) with a view of extending the research of Burnside and Dollar (2000), examined the impact of foreign aid in 46 sub-Saharan African countries using panel-based OLS regressions for the duration of 1996–2011. This study finds evidence that institutional quality and exports are positively significant, while inflation and M2 (money supply) were negatively significant. A policy index variable was included, and it showed a positive and significant effect on growth. Similarly, the aid variable displayed a largely positive and significant effect. However, when policy interacted with aid, the variable became insignificant which was contradictory to Burnside and Dollar (2000), who argue that the aid works better in a sound policy environment. In conclusion, the author reveals that the relationship between aid and growth is unclear and depends on specifications.

Furthermore, Hafiz et al. employed the dynamic panel estimation technique to study the impact of foreign aid on economic growth using evidence from a panel of South and East Asian countries for the period 1995–2013. The study showed significant and robust results that foreign aid promotes economic growth for the understudy countries, it further supports several empirical results on the positive and significant impact of external finance on domestic savings, investment, and growth, such as the study of Burnside and Dollar. They further conclude that foreign aid has a positive impact on economic growth in the underdeveloped economies if they have a good monetary, fiscal and trade policies but in the absence of good policies, there is a muted impact of aid on growth.

In addition, Moolio and Kong using panel data ranging from 1997 to 2014 and applying panel cointegration tests, panel fully modified ordinary least squares (FMOLS) and panel dynamic ordinary least squares (DOLS) estimated the magnitude of long-run relationship between aid and economic growth in Cambodia, Lao PDR, Myanmar and Vietnam. The results of panel cointegration tests indicate a robust, long-run relationship between the variables while panel FMOLS and panel DOLS estimation results show a positive impact of aid on economic growth, which generally concludes that foreign aid has a favorable impact on economic growth in the four countries.

More so, Osborne explores the impact of foreign aid on economic growth using variation in aid inflows from natural disasters. The study uses the disaster exposure of a country's "aid neighbors," defined as its competitors for aid from donors instead of using a country's own disaster exposure as an instrument for aid inflows which violates exogeneity assumptions; the study found no evidence of any long-run aid-growth effects, but further reveals that aid inflows significantly increase per capita GDP growth in the short to medium run due to increased household consumption while physical capital investment actually falls.

Galiani et al. (2014) while asserting that aid-growth literature has not found a convincing instrumental variable to identify the causal effects of aid exploits an instrumental variable because, since 1987, eligibility for aid from the International Development Association (IDA) has been based partly on whether or not a country is below a certain threshold of per capita income. Their paper finds evidence that other donors tend to reinforce rather than compensate for reductions in IDA aid following threshold crossings while focusing on the 35 countries that have crossed the income threshold from below between 1987 and 2010; a positive, statistically significant and economically sizable effect of aid on growth was found. The study further discovered a one percentage point increase in the aid to GNI ratio from the sample mean which raises annual real per capita growth in gross domestic product by approximately 0.35 percentage points. They concluded that increasing physical investment is the main channel through which aid promotes growth.

Khomba and Trew examine the local impact of foreign aid to constituencies and districts in Malawi over the period 1999 to 2013 using a highly detailed new aid database that includes annual disbursements at each project location. The study uses household panel survey which assumes that growth in light density is a good proxy for growth in per capita consumption as well as introducing a new political dataset that permits novel instrumental variables. The study further reveals that the impact on growth peaks after two to three years but then falls to zero which depicts that foreign aid has a level effect on incomes but does not stimulate sustained growth. In addition, the study concludes that bilateral aid appears to be better in causing growth than multilateral aid while aid delivered as a grant has an impact than aid given as a loan.

Feeny investigates the impact of foreign aid on economic growth in Papua New Guinea (PNG) using time-series data for the period 1965 to 1999 and employing the autoregressive distributed lag (ARDL) approach to cointegration proposed by Pesaran and Shin. The results show little or no evidence that aid and its various components have contributed to economic growth in Papua New Guinea. In addition, Juselius et al. examine the long-run effect of foreign aid (ODA) on key macroeconomic variables in 36 sub-Saharan African countries from the mid-1960s to 2007 using a well-specified cointegrated VAR model as a statistical benchmark.

Ali employed Johansen Cointegration test and vector error correction model (VECM) to examine the extent to which foreign aid to Egypt is effective especially in promoting economic growth; the study finds a negative and significant impact of foreign aid on economic growth in the long and short run. He further suggests that Egypt must rely upon the indigenous resources to promote development rather than depending on external factors.

Civelli et al. relied on the use of a spatial panel vector-autoregressive model (Sp P-VAR) for the analysis of sub-national effects of foreign aid on economic growth in Ugandan districts. Mapping nightlights to economic activity, the study found statistically significant positive and persistent effects of aid shocks on night-time luminosity, the results further suggest that the economic magnitude of these effects is small but significant with a multiplier in the long run.

Sahoo and Sethi used simple regression techniques to examine the impact of foreign aid on both economic growth and economic development of India from 1975–76 to 2009–10. The result shows that foreign aid has a significant positive impact on both economic growth and development of India; it further indicates that its impact on economic growth is higher than in economic development which means that aid contributes to economic growth but the growth is not translated into meaningful development in the long run which may be as a result of income inequality, poor economic policy, corruption, persistent mass poverty, faulty implementation of utilization strategy, underutilization of foreign aid or institutional inefficiency.

Foreign aid in Nigeria, as well as other developing countries, can be traced to the colonial era, while official development assistance became a prominent source of external financing after World War II because it was perceived as a solution to alleviate poverty and to better the economic situation of the developing countries. Fatukasi and Kudaisi observed that Western aid to Nigeria and other developing countries has been significant overtime amounting to $500 billion between 1990 and 1997. Against this backdrop, there have been studies in Nigeria and other countries to ascertain the impact of foreign aid to economic growth.

Methodological Frameworks to Study Aid

The study is based on the Harrod-Domar Model and Chenery and Strout's two-gap model because traditional pro-aid proponent advocated aid on the viewpoint that it complements domestic resources, eases foreign exchange constraints, transfers modern know-how and managerial skills and facilitates easy access to foreign markets there by contributing to economic growth (Okpanachi, 2011).

The Harrod-Domar Growth Model

The seminal paper by Sir Roy. F. Harrod in 1939 and Evsey Domar in 1946 which is generally referred to as the Harrod-Domar model of growth is the main theoretical framework used by most aid-growth literature. According to the model, physical capital formation (savings/investment) is a significant factor for achieving economic growth, hence, the incremental rate of output is equal to the savings rate divided by the incremental capital-output ratio.

$$\text{Ie } g = s / v$$

where "g" is the incremental rate of output, "s" is the savings rate and "v" is the incremental capital-output ratio, which implies that savings and growth are positively correlated connoting that they co-move in the same direction. In addition, the model introduces the Cobb-Douglas Production Function which stipulates that output is a function of capital and labor, given by $Y = f(K, L)$ where "Y" is output, "K" is capital and "L" is labor. Furthermore, most developing countries are usually labor intensive and are constrained by low-level capital which is a result of low savings; this lack of capital impeded growth of output since domestic savings are not sufficient to meet the required investment (capital accumulation) in less developed countries, due to their low per capita income, there is a need for foreign aid (Dollar & Easterly, 1999) to fill the savings-investment gap. The main objective of aid according to the Harrod-Domar model is to promote investment by augmenting savings. The new formula is given by:

$$g = (s + f) / v$$

where "f" is foreign aid. The extra savings, in the form of foreign aid, will enable a given economy to achieve a higher growth rate than what their domestic savings would have permitted. Over time, because of their potentially high savings rate, as assumed by the Keynesian theory the marginal propensity to save is greater than the average propensity to save, domestic resources become sufficient, and the need for foreign aid diminishes and eventually disappears (Panjak, 2005). Aid would then have fulfilled the aim of transforming the developing country from an aid-led development to self-sustaining development.

The Harrod-Domar model states that physical capital formation (savings/investment) is key to achieving economic growth, hence, the incremental rate of output is equal to the savings rate divided by the incremental capital-output ratio. Ie $g=s/v$ where "g" is the incremental rate of output, "s" is the savings rate and "v" is the incremental capital-output ratio, which implies that savings and growth are positively correlated connoting that they co-move in the same direction. In addition, the model introduces The Cobb-Douglas Production Function which stipulates that output is a function of capital and labor, given by $Y=f(K, L)$ Where "Y" is output, "K" is capital and "L" is labor. Furthermore, most developing countries are usually labor intensive and are constrained by low-level capital which is as a result of low savings, this lack of capital impeded growth of output since domestic savings are not sufficient to meet the required investment (capital accumulation) in less developed countries due to their low per capita income, there is a need for foreign aid (Dollar & Easterly, 1999) to fill the savings-investment gap. The main objective of aid according to the Harrod-Domar model is to promote investment by augmenting savings. The new formula is given by:

$$g = (s + f) / v$$

where "*f*" is foreign aid. The extra savings, in the form of foreign aid, will enable a given economy to achieve a higher growth rate than what their domestic savings would have permitted. Over time, because of their potentially high savings rate, as assumed by the Keynesian theory the marginal propensity to save is greater than the average propensity to save, domestic resources become enough, and the need for foreign aid diminishes and eventually disappears there by turning the country to a self-sustaining country (Panjak, 2005).

In Nigeria many studies have been undertaken in support of the positive impact of foreign aid on domestic savings which according to the Harrod-Domar model helps a country to make up for her insufficient funds and supplements domestic savings, thereby allowing such a country to experience a higher rate of investment which, in turn, promotes faster growth. For instance, Bakare et al. (2014) while testing the validity of Harrod-Domar model in Nigeria using the ordinary least square multiple regression analytical method observed that there is a significant relationship between capital formation and economic growth in Nigeria.

In addition, Orji et al. while evaluating the transmission link through which foreign aid transmits to affect economic growth using the autoregressive distributed lag (ARDL) model finds that foreign aid has a positive and significant impact on capital formation in Nigeria.

Two-Gap Model

The two-gap model also known as the dual-gap model of economic development as popularized by Hollis Chenery and Alan Strout is an extension of the Harrod-Domar model. Chenery and Strout (1966) argue that the development of less developed countries is constrained due to the presence of two gaps:

1. The gap between domestic savings and investment, where domestic savings are inadequate to support the level of growth.
2. Gap between export revenues and imports, or foreign exchange gap, where import purchasing power (value of imports + capital transfers) is inadequate to support the level of growth.

Akande while exploring the importance and application of the theoretical prescriptions of the two-gap model to the Nigerian economic growth agreed that the model predicts that foreign aid and foreign direct investment are an optimal means to breaking the poverty circle and solving the two gaps simultaneously. He further stated that the vicious circle of poverty in Nigerian economy cannot be broken by attracting foreign aid and foreign direct investment alone except if some fundamental rigidities which include, among other things, inappropriate economic policies, corruption, mismanagement of resources and overreliance on oil resources are eliminated.

In addition, Aremu et al. in their study while trying to unveil the existence of the two-gaps in the Nigerian economy discovered the existence of both savings gap and exchange rate gap because domestic savings are insufficient to fund required investment in Nigeria as well as disequilibrium in external balance which retards economic performance. They further noted that although FDI acts as a bridge, it is insufficient in the short run and not reliable in the long run as it promotes importation in both periods, which could widen the existing exchange rate gap.

The trade gap exists because, in most LDCs, export earnings are way below import requirements, as a result, there is a deficit of foreign exchange to finance the importation of capital and intermediate goods necessary for the manufacturing of investment goods. Therefore, in order to achieve a sustainable growth level, there is a need for foreign assistance in the form of aid (Delessa, 2012).

Three-Gap Model

Edmar Bacha and Lance Taylor (1994) in their studies made an extension of the two-gap model with an inclusion of a third gap called the fiscal gap. According to them, it occurs when government expenditure exceeds government revenue which leads to budget deficit. More so, because of the underdeveloped tax system and the institutional weaknesses of developing countries, generated revenue is insufficient to finance necessary recurrent and capital expenditure, with a view of financing the budget deficit, the government may consider borrowing from either the private sector or the central bank. If it is the former, low per capita income of the population makes it impossible for domestic savings to fill this gap, while in the latter, the risk of nominal inflation keeps this option closed. The model recommends foreign aid as a key option in bridging the fiscal deficit gap (Delessa, 2012).

Functional Relationship and Model Specification

Different studies have employed different variables and methodology to analyze the impact of foreign aid on economic growth in developing countries, while some studies focused on the impact of foreign aid on savings and investment like Okpanachi (2011), others like Nwosu (2018), Mustafa et al. (2018), Yiew and Lau (2018), Siddique and Kiani (2017), Fasanya and Onakoya (2012), Okoro et al. (2019), Tang and Bundhoo (2017), Liew et al. (2012), Mbah and Amassoma (2014), and Kolawole (2013), focused on the impact of foreign aid on economic growth while others carried out sectors-based impact analysis for Kumi, Selaya and Chukwuemeka. The major objective of this research is to examine the impact of foreign aid on the economic growth of Nigeria. Therefore, in line with the theoretical framework, the variables employed include gross domestic product, used as a proxy for economic growth, aid (Overseas Development Assistance, ODA). Investment (foreign direct investment), export and import will be used in the estimation which will be given as: GDP = f(AID, INV, EXP, IMP).

Econometric Estimation:

$$\text{GDP}_t = b_0 + b_1\text{LAID}_t + b_2\text{LINV}_t + b_3\text{LEXP}_t + b_4\text{LIMP}_t + U_t$$

Where GDP = Gross Domestic Product.
AID = Aid in the form of ODA
INV = Investment
EXP = Export
IMP = Import

$b1, \ldots, b4$ are coefficients, U is the error term while t represents time.

Sources of Data

The success of any econometric analysis ultimately depends on the availability and accuracy of data, for reliability the annual time series data will be generated from the Central Bank of Nigeria Statistical Bulletin and World Bank World Development Indicators.

Data Analysis and Estimation Techniques

Since the study requires time series data, some pre-test must be executed before

estimation can take place, which includes unit root tests (which is used to check for the stationarity of the variables), this test will be conducted using Augmented Dickey-Fuller (ADF), this enables us to test for stationarity of the variables included in the model, and a cointegration test (to establish if the variables exhibit a long-run relationship) is conducted.

This study examines the impact of foreign aid on the economic growth of Nigeria from 1981–2017 using the annual time series data generated from the Central Bank of Nigeria statistical bulletin and World Bank indicators. The study follows the theoretical framework of the two-gap model of economic growth coupled with an assumption of open economy. In addition, the study employed various econometric techniques.

Therefore, in line with the theoretical framework of the two-gap model of economic growth coupled with an assumption of open economy, the functional relationship between foreign aid and economic growth of Nigeria is expressed thus:

$$\text{LGDP} = f\left(\text{LAID},\text{LINV},\text{LEXP},\text{LIMP}\right)\ldots\ldots \quad (6.1)$$

The variables are in natural log, this will help to improve the linearity of the parameters and to avoid heteroscedasticity. (Nwosu, 2018).

Therefore in a linear function Eq. (6.1) can be represented as

$$\text{LGDP}_t = b_0 + b_1\text{LAID}_t + b_2\text{LINV}_t + b_3\text{LEXP}_t + b_4\text{LIMP}_t + U_t\ldots\ldots \quad (6.2)$$

Where GDP = Gross Domestic Product
LAID = aid in the form of ODA

LINV = Investment
LEXP = export
LIMP = import
*b*1,…, *b*4 are coefficients, *U* is the error term while *t* represents time.

The Johansen test of cointegration is used in this analysis (Mbah & Amassona, 2014). Also, in consonance with economic theory, it is expected that b2 and b3 are positive, b1 can either be positive or negative and b4 is negative.

Unit Root Test

A time series is said to be stationary if its mean and variance are constant over time, and the series has a theoretical correlogram that diminishes as the lag length increases. Stationarity in time series variables is necessary because unless a non-stationary series is differenced, any regressions carried out will produce spurious (non-meaningful) results (Nwosu, 2018). The unit root test employed in this study to check for the stationarity of the variable is the Augmented Dickey-Fuller (ADF) test.

To test for stationarity or the absence of unit roots, this test is carried out using the Augmented Dickey-Fuller test procedure. Under the stationarity test, we take the following decision rule: if the absolute value of the Augmented Dickey-Fuller (ADF) test is greater than the critical value either at 1%, 5% or 10% level of significance at the order zero, one, or two, it shows that variable under consideration is stationary otherwise it is not.

The results reported in Table 6.1 confirm the acceptance of the null hypothesis of unit root at all level of significance for each variable based on the ADF test. From the table, it could be seen that first differencing of LGDP and LAID variables yields rejection of the null hypothesis on unit root at 5% level of significance, while LINV, LEXP and LIMP were stationary at level form. Based on these test results, it concludes that all the variables that is LGDP, LAID, LINV, LEXP and LIMP are stationary which implies that the combination of one or more of these series may exhibit a long-run relationship.

Cointegration Test

Table 6.2 presents results from the Johansen cointegration analysis. Evidence from the above cointegration test depicts that the trace value statistics reveal that the null hypothesis of no cointegrating vector is being rejected at the 5% level of significance. The result further showed that there are three cointegrating vectors among the variables of interest at the

Table 6.1 The result of unit root test

Variables	*Unit root test*		
	Augmented dickey-fuller (ADF) test		
	Test statistic	*5% critical value*	*Decision*
LRGDP	-2.106**	-1.708	1(1)
LAID	-3.031***	-1.708	1(1)
LINV	-2.360**	-1.706	1(0)
LEXP	-2.510***	-1.706	1(0)
LIMP	-2.912***	-1.706	1(0)

Data Source: CBN Statistical Bulletin and World Development Indicator

Note: One, two and three asterisks denote rejection of the null hypothesis at 1%, 5% and 10% respectively based on critical values

Table 6.2 The result of the Johansen cointegration test

Johansen cointegration test				
Maximum rank	*Trace statistic*	*5% Critical value*	*Max statistic*	*5% Critical value*
0	111.9905	68.52	48.0003	33.46
1	63.9902	47.21	25.5548	27.07
2	38.4354	29.68	24.5074	20.97
3	13.9280*	15.41	11.6234	14.07
4	2.3046	3.76	2.3046	3.76

Note (*) denotes rejection of the null hypothesis at the 5% level of significance and maximum rank represents the number of cointegrating equations

5% level of significance, which in turn stipulates that there exists a long-run relationship among the variables that is GDP, foreign aid, investment, export and import.

The result in Table 6.3 shows that the first lag of LAID has a short-run causal effect on LGDP at a significant level of 5% on average ceteris paribus, while second lag, third lag and fourth lag of LAID have no short-run impact on LGDP ceteris paribus.

For investment and export variable, it shows that first lag of LINV, fourth lag of LINV, first lag of LEXP and third lag of LEXP have a short-run causal and significant effect on LRGDP while lag two of LINV, lag three of LINV, lag two of LEXP and lag four of LEXP have no short-run causal effect on LRGDP.

Table 6.3 The result of the short-run relationship

LGDP	*Coefficient*	*Probability*
LGDP		
L1.	0.7520928	0.000
L2.	0.1431093	0.469
L3.	-0.1486234	0.428
L4.	0.0653822	0.631
LAID		
L1.	-0.0214872	0.048
L2.	0.0011227	0.931
L3.	-0.0149362	0.172
L4.	0.0243665	0.035
LINV		
L1.	-0.0339	0.011
L2.	0.0017528	0.849
L3.	-0.0149496	0.179
L4.	-0.0735996	0.000
LEXP		
L1.	0.1110666	0.001
L2.	-0.0143185	0.581
L3.	0.0761056	0.033
L4.	-0.0477373	0.068
LIMP		
L1.	-0.0462536	0.129
L2.	0.0363979	0.207
L3.	-0.0444218	0.183
L4.	-0.0062909	0.790
_CONS	2.054853	0.008

Source: CBN Statistical Bulletin and World Development Indicator

In addition, all the lags of variable show that import has no short-run causal effect on gross domestic product ceteris paribus.

Table 6.4 shows that there is a bidirectional causality from: LAID to LGDP, LINV to LGDP, LEXP to LGDP and LIMP to LINV vice-versa. In addition, there is also a unidirectional causality from LGDP to LIMP, LIMP to LEXP, LEXP to LAID and LINV to LAID. In summary, it shows that on the short-run LAID Granger causes LGDP vice-versa, LINV Granger causes LGDP vice-versa, LEXP Granger causes LGDP vice-versa and LIMP Granger causes LINV vice-versa.

Table 6.4 The result of the Granger causality Wald test

Granger Causality Wald Tests

Equation	*Excluded*	*F*	*Probability*
LGDP	LAID	5.576	0.0090
LGDP	LINV	10.975	0.0006
LGDP	LEXP	6.6697	0.0046
LGDP	LIMP	1.2389	0.3460
LGDP	ALL	5.4712	0.0025
LAID	LGDP	5.0615	0.0127
LAID	LINV	0.68599	0.6153
LAID	LEXP	3.7787	0.0326
LAID	LIMP	0.76816	0.5661
LAID	ALL	3.9648	0.0101
LINV	LGDP	6.4978	0.0051
LINV	LAID	3.3506	0.0463
LINV	LEXP	1.5754	0.2436
LINV	LIMP	8.0878	0.0021
LINV	ALL	6.5315	0.0011
LEXP	LGDP	4.2426	0.0228
LEXP	LAID	1.225	0.3511
LEXP	LINV	10.428	0.0007
LEXP	LIMP	4.8869	0.0143
LEXP	ALL	5.4913	0.0024
LIMP	LGDP	4.0995	0.0254
LIMP	LAID	0.98449	0.4522
LIMP	LINV	5.3326	0.0105
LIMP	LEXP	1.3129	0.3201
LIMP	ALL	3.1977	0.0237

Ordinary Least Square

Some studies on aid-growth relationship in Nigeria treated aid as an exogenous variable (see Mbah & Amassoma, 2014; Kolawole, 2013; Ugwuebe et al.; Nwosu, 2018) while others have employed the use of instruments in order to address the issue of reverse causality between foreign aid at the level of per capital income/output (see Olanrele & Ibrahim, 2015. The estimating equation is (Table 6.5):

$$\text{LGDP}_t = b_0 + b_1\text{LAID}_t + b_2\text{LINV}_t + b_3\text{LEXP}_t + b_4\text{LIMP}_t + U_t \ldots\ldots \quad (6.3)$$

Table 6.3 depicts the result of long-run impact of foreign aid on economic growth. The result indicates that the coefficient of determination is

Table 6.5 Ordinary least squares regression result

Dependent variable: LGDP

Variables	*Coefficient*	*Standard Error*	*t-Statistic*	*Probability*
LAID	0.1551162	0.0282694	5.49	0.000
LINV	-0.155801	0.0325821	-4.78	0.013
LEXP	-0.1043827	0.0754542	-1.38	0.176
LIMP	0.2336607	0.071953	3.25	0.003
CONS	9.645454	0.2105287	45.82	0.000

R-squared 0.9477 F-stat 147.02
Adjusted R-squared 0.9412 Prob 0.0000
Durbin-Watson d-statistic (5, 37) = 1.147436

Data Source: CBN Statistical Bulletin and World Development Indicator

0.9477 which signifies that over 94% variation in GDP is accounted for by the explanatory variables in the model. The prob. (F-statistic) of 0.001 shows an overall goodness of fit of the model. The Durbin Watson (D.W) statistics of 1.1473 shows the absence of auto-correlation or first-order serial correlation. In terms of the signs of the coefficients, not all the variables concur with a priori theoretical expectation.

Aid is positive and statistically significant with coefficient of 0.1551, this indicates that an increase of 1% in aid leads to an economic growth increase by 0.1551%, this further implies that aid contributes to economic growth in Nigeria, this follows the apriori expectation as supported by the two-gap model by Chenery and Strout (1966) and Delessa (2012) who noted that for a less developing country to achieve a sustainable growth level there is need for foreign aid. These results authenticate the findings of Nwosu (2018), Fasanya and Onakoya, Delessa (2012), Papanek and Karras that foreign aid has a substantially greater effect on economic growth. Gitaru noted that aid contribution to economic growth can be indirect and direct especially through the form of budget support. The result also indicates that investment did not follow theoretical expectation because the sign is negative, this can be as a result that FDI is used to facilitate the importation of consumer goods instead of capital goods. For Nigeria to experience the gains of foreign direct investment Olatunji and Shahid suggested the need to improve the business environment and the provision of necessary infrastructure.

Furthermore, export is negative and non-significant. This did not follow the apriori expectation, it also contradicts the export-led growth hypothesis which stipulates that export is a key driver of growth. However, in an undiversified and a mono-economy like Nigeria, this is not surprising because since 1960 oil accounts for about 80% of total export and foreign exchange earnings of about $300 billion dollars, that has not translated to economic growth and development of Nigeria.

In addition, import is positive and statistically significant defiling the apriori expectation, but the results justify that Nigeria is an import dependent economy.

Conclusion and Policy Recommendations

The study analyzes the impact of aid on economic growth in Nigeria within the period of 1981–2017. This impact is examined using the ordinary least square (OLS) and Johansen cointegration test to test for the correlation between the variables and to also determine the long-run linear relationship among the variables of interest. In addition, Granger causality Wald test and T-statistics were used to determine the short-run effects. The estimates of the long-run approaches to cointegration agree that foreign aid has a positive impact on economic growth in Nigeria, which indicates that foreign aid has a favorable effect on Nigeria's economic growth. The result of this study is consistent with findings from similar studies on aid-growth relationship in Nigeria such as Nwosu (2018), Olanrele and Ibrahim (2015), Fasanya and Onakoya (2012), and Reddy and Minoiu. In addition, the estimates of Granger causality Wald test depict that on the short run foreign aid has a significant effect on economic growth in Nigeria.

The result equally shows that investment negatively impacts on the economy, indicating that business climate is not conducive for foreign investment in Nigeria. The export variable equally indicates a negative sign both in the short and long run, implying that export has no impact on economic growth in Nigeria, while import, on the other hand, showed a positive impact on the economic growth in Nigeria.

Conclusively, the evidence of this study provides the support that foreign aid is vital in the promotion of a country's economic development. In Nigeria's case, foreign aid seems not to have been in vain, although the magnitude of the effect has not been sufficiently strong to reduce the level of poverty. Aid itself does not promote economic growth but a country's economic growth depends to a large extent on sound economic policies (Burnside & Dollar, 2000). Amid the huge aid inflow, Nigeria's economic growth is still characterized by high unemployment rate, double-digit inflation and high poverty rate which according to the World Poverty Clock estimate is expected to increase from 44.2% to 45.5% in 2030.

Therefore, for Nigeria to reap the benefits of foreign aid, there is need to focus more on the implementation of sound macroeconomic policies. For example, industrial policy, policies that strengthen institutional weaknesses and good governance improve business and regulatory environment, diversification of her economy providing the necessary infrastructure.

Appendix 1: Data Set

Year	*LIMP*	*LEXP*	*LAID*	*LINV*	*LGDP*
1981	1.108551	1.042312	7.962322	-0.48184	11.11858
1982	1.032236	0.914153	7.913178	-0.52055	11.08798
1983	0.949571	0.875206	8.034428	-0.42558	11.03774
1984	0.856022	0.958468	7.910678	-0.58935	11.03287
1985	0.848965	1.068957	7.89603	-0.18148	11.05782
1986	0.776963	0.950394	8.039255	-0.45279	11.05808
1987	1.251923	1.48231	8.026492	0.06411	11.07176
1988	1.33134	1.494054	8.20828	-0.11765	11.1025
1989	1.489399	1.763212	8.680272	0.631656	11.11076
1990	1.660086	2.040943	8.539766	0.036609	11.15911
1991	1.951766	2.084703	8.538611	0.161463	11.16066
1992	2.155795	2.313048	8.512204	0.273237	11.18032
1993	2.219137	2.339988	8.592865	0.685544	11.1714
1994	2.211625	2.313992	8.396896	0.762742	11.16344
1995	2.87802	2.978026	8.374345	0.389062	11.16313
1996	2.75022	3.11712	8.355375	0.494126	11.18098
1997	2.927225	3.094004	8.406489	0.451304	11.19355
1998	2.922943	2.876135	8.422557	0.284513	11.20462
1999	2.935767	3.075171	8.287533	0.228544	11.20715
2000	2.993446	3.289081	8.365357	0.215304	11.2284
2001	3.132957	3.271366	8.366273	0.206363	11.25337
2002	3.179751	3.241591	8.599807	0.293302	11.31531
2003	3.318112	3.489661	8.572081	0.281366	11.3461
2004	3.298208	3.66302	8.798029	0.138014	11.38452

(*continued*)

(continued)

Year	*LIMP*	*LEXP*	*LAID*	*LINV*	*LGDP*
2005	3.447291	3.86013	9.804302	0.451607	11.41162
2006	3.492554	3.864789	10.05223	0.313028	11.43717
2007	3.592394	3.919588	9.262534	0.340431	11.46489
2008	3.747659	4.016519	9.09135	0.3859	11.49332
2009	3.738833	3.934817	9.2034	0.467002	11.52689
2010	3.911902	4.079596	9.299272	0.219709	11.56034
2011	4.041229	4.18289	9.223104	0.333369	11.5828
2012	3.989741	4.180107	9.251424	0.187247	11.60079
2013	3.974946	4.183612	9.367194	0.03352	11.62884
2014	4.02279	4.112622	9.352721	-0.08714	11.65541
2015	4.044386	3.946706	9.380481	-0.19768	11.66678
2016	3.976825	3.946237	9.397625	0.040803	11.6597
2017	4.033619	4.14576	9.519638	-0.03117	11.66319

Source: World Development Indicator, Central Bank of Nigeria Statistical Bulletin

Appendix 2: OLS Regression Table

. **reg LGDP LAID LINV LEXP LIMP**

Number of obs	=	37
F(4, 32)	=	145.02
Prob > F	=	0.0000
R-squared	=	0.9477
Root MSE	=	0.05182

Source	SS	df	MS
Model	1.5579094	4	0.38947735
Residual	0.085943047	32	0.00268572
Total	1.64385245	36	0.045662568

Adj R-squared = 0.9412

LGDP	Coef.	Std. Err.	t	P>|t|	[95% Conf.	Interval]
LAID	0.1551162	0.0282694	5.49	0.000	0.0975333	0.212699
LINV	-0.155801	0.0325821	-4.78	0.000	-0.2221685	-0.0894335
LEXP	-0.1043827	0.0754542	-1.38	0.176	-0.2580779	0.0493124
LIMP	0.2336607	0.071953	3.25	0.003	0.0870973	0.3802241
_cons	9.645454	0.2105287	45.82	0.000	9.216621	10.07429

45 . estat dwatson

Durbin-Watson d-statistic(5, 37) = 1.147436

Appendix 3: Optimal Lag Length Table

6 . varsoc lgdp, maxlag(10) exog(laid linv lexp limp)

Selection-order criteria

Sample: 1991 - 2017 Number of obs = 27

lag	LL	LR	df	p	FPE	AIC	HQIC	SBIC
0	44.8612				0.00307	-2.95268	-2.88132	-2.71271
1	80.9287	72.135	1	0.000	0.000229	-5.55028	-5.46465	-5.26231
2	83.7335	5.6095	1	0.018	0.000201	-5.68396	-5.58406	-5.348
3	86.5421	5.6174	1	0.018	0.000177	-5.81794	-5.70377	-5.43399
4	88.2755	3.4667	1	0.063	.000169*	-5.87226*	-5.74382*	-5.44031*
5	89.1257	1.7003	1	0.192	0.000173	-5.86116	-5.71845	-5.38122
6	89.8846	1.5179	1	0.218	0.000178	-5.8433	-5.68632	-5.31537
7	89.9375	0.10574	1	0.745	0.000195	-5.77315	-5.60189	-5.19722
8	90.6444	1.4139	1	0.234	0.000203	-5.75144	-5.56591	-5.12752
9	90.7372	0.18544	1	0.667	0.000223	-5.68423	-5.48444	-5.01232
10	93.5738	5.6733*	1	0.017	0.0002	-5.82028	-5.60621	-5.10037

Endogenous: lgdp

Exogenous: laid linv lexp limp _cons

Appendix 4: Augmented Dickey-Fuller Test Table

LGDP

8 . dfuller LGDP, trend lags(4)			
Augmented Dickey-Fuller test for unit root		Number of obs =	32
Interpolated Dickey-Fuller			
Test	1% Critical	5% Critical	10% Critical
Statistic	Value	Value	Value
Z(t) -2.235	-4.316	-3.572	-3.223
MacKinnon approximate p-value for Z(t) = **0.4700**			

9 . dfuller LGDP, drift lags(4)			
Augmented Dickey-Fuller test for unit root		Number of obs =	32
Z(t) has t-distribution			
Test	1% Critical	5% Critical	10% Critical
Statistic	Value	Value	Value
Z(t) -0.471	-2.479	-1.706	-1.315
p-value for Z(t) = **0.3208**			

MacKinnon approximate p-value for Z(t) = **0.7510**
15 . dfuller DLGDP, drift lags(4)
Augmented Dickey-Fuller test for unit root Number of obs = 31

	Z(t) has t-distribution Test Value	1% Critical Value	5% Critical Value	10% Critical Statistic
Z(t)	-2.106	-2.485	-1.708	-1.316

p-value for Z(t) = **0.0227**

LAID

18	. dfuller	LAID, trend lags(4)				
	Augmented	Dickey-Fuller test	for unit root	Number of	obs	= 32

	Interpolated Dickey-Fuller Test Value	1% Critical Value	5% Critical Value	10% Critical Statistic
Z(t)	-2.639	-4.316	-3.572	-3.223

MacKinnon approximate p-value for Z(t) = **0.2622**
19 . dfuller LAID, drift lags(4)
Augmented Dickey-Fuller test for unit root Number of obs = 32

	Z(t) has t-distribution Test Value	1% Critical Value	5% Critical Value	10% Critical Statistic
Z(t)	-1.086	-2.479	-1.706	-1.315

(*continued*)

(continued)

21 . dfuller DLAID, drift lags(4)
Augmented Dickey-Fuller test for unit root Number of obs = 31

Z(t) has t-distribution

	Test Value	1% Critical Value	5% Critical Value	10% Critical Statistic
Z(t)	**-3.031**	**-2.485**	**-1.708**	**-1.316**

p-value for Z(t) = **0.0028**

LINV

22 . dfuller LINV, drift lags(4)
Augmented Dickey-Fuller test for unit root Number of obs = **32**

Z(t) has t-distribution

	Test Statistic	1% Critical Value	5% Critical Value	10% Critical Value
Z(t)	**-2.360**	**-2.479**	**-1.706**	**-1.315**

p-value for Z(t) = **0.0130**

LEXP

24 . dfuller LEXP, drift lags(4)
Augmented Dickey-Fuller test for unit root Number of obs = **32**

Z(t) has t-distribution

	Test Value	1% Critical Value	5% Critical Value	10% Critical Statistic
Z(t)	**-2.510**	**-2.479**	**-1.706**	**-1.315**

p-value for Z(t) = **0.0093**

LIMP

26 . dfuller LIMP, drift lags(4)
Augmented Dickey-Fuller test for unit root Number of obs = **32**

Z(t) has t-distribution

	Test Statistic	1% Critical Value	5% Critical Value	10% Critical Value
Z(t)	**-2.912**	**-2.479**	**-1.706**	**-1.315**

p-value for Z(t) = **0.0036**

Appendix 5: Johansen Cointegration Test

7 . vecrank LGDP LAID LINV LEXP LIMP, trend(constant) lags(4) max

Johansen tests for cointegration

Trend: constant Number of obs = 33

Sample: 5 - 37 Lags = 4

maximum rank	parms	LL	eigenvalue	trace statistic	5% critical value
0	80	207.95673	.	111.9905	68.52
1	89	231.95687	0.76650	63.9902	47.21
2	96	244.73428	0.53901	38.4354	29.68
3	101	256.98797	0.52415	13.9280*	15.41
4	104	262.79966	0.29688	2.3046	3.76
5	105	263.95198	0.06745		

maximum rank	parms	LL	eigenvalue	max statistic	5% critical value
0	80	207.95673	.	48.0003	33.46
1	89	231.95687	0.76650	25.5548	27.07
2	96	244.73428	0.53901	24.5074	20.97
3	101	256.98797	0.52415	11.6234	14.07
4	104	262.79966	0.29688	2.3046	3.76
5	105	263.95198	0.06745		

Appendix 6: Granger Causality Wald Test

8 . vargranger

Granger causality Wald tests

Equation	Excluded	F	df	df_r	Prob > F
lgdp	laid	5.576	4	12	0.0090
lgdp	linv	10.975	4	12	0.0006
lgdp	lexp	6.6697	4	12	0.0046
lgdp	limp	1.2389	4	12	0.3460
lgdp	ALL	5.4712	16	12	0.0025
laid	lgdp	5.0615	4	12	0.0127
laid	linv	0.68599	4	12	0.6153
laid	lexp	3.7787	4	12	0.0326
laid	limp	0.76816	4	12	0.5661
laid	ALL	3.9648	16	12	0.0101
linv	lgdp	6.4978	4	12	0.0051
linv	laid	3.3506	4	12	0.0463
linv	lexp	1.5754	4	12	0.2436
linv	limp	8.0878	4	12	0.0021
linv	ALL	6.5315	16	12	0.0011
lexp	lgdp	4.2426	4	12	0.0228
lexp	laid	1.225	4	12	0.3511
lexp	linv	10.428	4	12	0.0007
lexp	limp	4.8869	4	12	0.0143

(*continued*)

(continued)

lexp	ALL	5.4913	16	12	0.0024
limp	lgdp	4.0995	4	12	0.0254
limp	laid	0.98449	4	12	0.4522
limp	linv	5.3326	4	12	0.0105
limp	lexp	1.3129	4	12	0.3201
limp	ALL	3.1977	16	12	0.0237

Appendix 7: Short Run t-test

7 . var lgdp laid linv lexp limp, lags(1/4) small						
Vector autoregression						
Sample: 1985 - 2017				Number of obs	=	33
Log likelihood =	263.9522			AIC	=	-9.633464
FPE =	1.43E-10			HQIC	=	-8.031326
Det(Sigma_ml) =	7.77E-14			SBIC	=	-4.87185
Equation	Parms	RMSE	R-sq	F	P > F	
lgdp	21	0.012713	0.9986	1177.867	0.0000	
laid	21	0.213154	0.9438	27.70215	0.0000	
linv	21	0.152988	0.8740	11.44057	0.0001	
lexp	21	0.140433	0.9925	219.452	0.0000	
limp	21	0.140796	0.9927	224.0262	0.0000	

		Coef.	Std. Err.	t	P>|t|	[95% Conf.	Interval]
lgdp							
	lgdp						
	L1.	0.7520928	0.1451197	5.18	0.000	0.435904	1.068281
	L2.	0.1431093	0.1915702	0.75	0.469	-0.2742862	0.5605049
	L3.	-0.1486234	0.1812456	-0.82	0.428	-0.5435237	0.2462769
	L4.	0.0653822	0.1327637	0.49	0.631	-0.2238851	0.3546495
	laid						
	L1.	-0.0214872	0.0097685	-2.20	0.048	-0.0427709	-0.0002034
	L2.	0.0011227	0.0126956	0.09	0.931	-0.0265387	0.028784
	L3.	-0.0149362	0.0102822	-1.45	0.172	-0.0373393	0.0074669
	L4.	0.0243665	0.0102611	2.37	0.035	0.0020096	0.0467235
	linv						
	L1.	-0.0339	0.0113635	-2.98	0.011	-0.058659	-0.009141
	L2.	0.0017528	0.0090301	0.19	0.849	-0.0179221	0.0214278
	L3.	-0.0149496	0.0104696	-1.43	0.179	-0.0377608	0.0078617
	L4.	-0.0735996	0.0116763	-6.30	0.000	-0.0990401	-0.0481591
	lexp						
	L1.	0.1110666	0.0256569	4.33	0.001	0.0551649	0.1669682
	L2.	-0.0143185	0.0252701	-0.57	0.581	-0.0693773	0.0407403
	L3.	0.0761056	0.0315727	2.41	0.033	0.0073145	0.1448967
	L4.	-0.0477373	0.0238062	-2.01	0.068	-0.0996065	0.004132
	limp						
	L1.	-0.0462536	0.028373	-1.63	0.129	-0.108073	0.0155658
	L2.	0.0363979	0.0273057	1.33	0.207	-0.0230961	0.0958919
	L3.	-0.0444218	0.0314229	-1.41	0.183	-0.1128864	0.0240427
	L4.	-0.0062909	0.0230597	-0.27	0.790	-0.0565337	0.0439518
	_cons	2.054853	0.6470136	3.18	0.008	0.6451319	3.464575

(continued)

(continued)

7 . var lgdp laid linv lexp limp, lags(1/4) small						
Vector autoregression						
Sample: 1985 - 2017				Number of obs	=	33
Log likelihood =	263.9522			AIC	=	-9.633464
FPE =	1.43E-10			HQIC	=	-8.031326
Det(Sigma_ml) =	7.77E-14			SBIC	=	-4.87185
Equation	Parms	RMSE	R-sq	F	P > F	
lgdp	21	0.012713	0.9986	1177.867	0.0000	
laid	21	0.213154	0.9438	27.70215	0.0000	
linv	21	0.152988	0.8740	11.44057	0.0001	
lexp	21	0.140433	0.9925	219.452	0.0000	
limp	21	0.140796	0.9927	224.0262	0.0000	

	Coef.	Std. Err.	t	P>|t|	[95% Conf.	Interval]
lgdp						
lgdp						
L1.	0.7520928	0.1451197	5.18	0.000	0.435904	1.068281
L2.	0.1431093	0.1915702	0.75	0.469	-0.2742862	0.5605049
L3.	-0.1486234	0.1812456	-0.82	0.428	-0.5435237	0.2462769
L4.	0.0653822	0.1327637	0.49	0.631	-0.2238851	0.3546495
laid						
L1.	-0.0214872	0.0097685	-2.20	0.048	-0.0427709	-0.0002034
L2.	0.0011227	0.0126956	0.09	0.931	-0.0265387	0.028784
L3.	-0.0149362	0.0102822	-1.45	0.172	-0.0373393	0.0074669
L4.	0.0243665	0.0102611	2.37	0.035	0.0020096	0.0467235
linv						
L1.	-0.0339	0.0113635	-2.98	0.011	-0.058659	-0.009141
L2.	0.0017528	0.0090301	0.19	0.849	-0.0179221	0.0214278
L3.	-0.0149496	0.0104696	-1.43	0.179	-0.0377608	0.0078617
L4.	-0.0735996	0.0116763	-6.30	0.000	-0.0990401	-0.0481591
lexp						
L1.	0.1110666	0.0256569	4.33	0.001	0.0551649	0.1669682
L2.	-0.0143185	0.0252701	-0.57	0.581	-0.0693773	0.0407403
L3.	0.0761056	0.0315727	2.41	0.033	0.0073145	0.1448967
L4.	-0.0477373	0.0238062	-2.01	0.068	-0.0996065	0.004132
limp						
L1.	-0.0462536	0.028373	-1.63	0.129	-0.108073	0.0155658
L2.	0.0363979	0.0273057	1.33	0.207	-0.0230961	0.0958919
L3.	-0.0444218	0.0314229	-1.41	0.183	-0.1128864	0.0240427
L4.	-0.0062909	0.0230597	-0.27	0.790	-0.0565337	0.0439518
_cons	2.054853	0.6470136	3.18	0.008	0.6451319	3.464575
laid						
lgdp						
L1.	1.312975	2.433258	0.54	0.599	-3.98864	6.614589
L2.	-0.6311592	3.212105	-0.20	0.848	-7.629734	6.367416
L3.	7.705785	3.03899	2.54	0.026	1.084394	14.32718
L4.	-8.130947	2.226082	-3.65	0.003	-12.98116	-3.280731
laid						
L1.	0.760616	0.163791	4.64	0.001	0.4037461	1.117486
L2.	-0.6420996	0.2128705	-3.02	0.011	-1.105905	-0.1782947
L3.	0.5223202	0.1724049	3.03	0.010	0.1466822	0.8979582
L4.	-0.111361	0.1720497	-0.65	0.530	-0.486225	0.263503

(*continued*)

(continued)

	linv						
	L1.	-0.1296164	0.1905348	-0.68	0.509	-0.5447561	0.2855233
	L2.	0.0087821	0.1514105	0.06	0.955	-0.321113	0.3386773
	L3.	-0.1347684	0.1755463	-0.77	0.458	-0.5172508	0.2477141
	L4.	-0.2773434	0.1957795	-1.42	0.182	-0.7039104	0.1492235
	lexp						
	L1.	0.3975606	0.4301962	0.92	0.374	-0.5397563	1.334878
	L2.	0.1091503	0.4237097	0.26	0.801	-0.8140338	1.032334
	L3.	-0.8373068	0.5293881	-1.58	0.140	-1.990744	0.3161308
	L4.	-0.3106974	0.3991643	-0.78	0.451	-1.180402	0.5590069
	limp						
	L1.	-0.0909732	0.4757367	-0.19	0.852	-1.127514	0.9455681
	L2.	0.2834556	0.4578412	0.62	0.547	-0.7140946	1.281006
	L3.	0.5542536	0.526875	1.05	0.314	-0.5937085	1.702216
	L4.	0.0885641	0.3866474	0.23	0.823	-0.7538682	0.9309963
	_cons	0.6903175	10.84864	0.06	0.950	-22.94683	24.32747
linv							
	lgdp						
	L1.	-3.786276	1.746434	-2.17	0.051	-7.591429	0.0188765
	L2.	0.5750634	2.305439	0.25	0.807	-4.448056	5.598183
	L3.	4.450823	2.181189	2.04	0.064	-0.3015785	9.203225
	L4.	-4.368995	1.597736	-2.73	0.018	-7.850163	-0.8878266
	laid						
	L1.	0.3259924	0.1175585	2.77	0.017	0.0698545	0.5821303
	L2.	-0.2081653	0.1527845	-1.36	0.198	-0.5410541	0.1247236
	L3.	0.393947	0.123741	3.18	0.008	0.1243387	0.6635554
	L4.	0.0043148	0.123486	0.03	0.973	-0.2647381	0.2733677
	linv						
	L1.	-0.0988622	0.1367534	-0.72	0.484	-0.3968224	0.1990979
	L2.	0.1919542	0.1086726	1.77	0.103	-0.0448229	0.4287314
	L3.	-0.3711133	0.1259956	-2.95	0.012	-0.6456342	-0.0965924
	L4.	-0.1096368	0.1405177	-0.78	0.450	-0.4157986	0.1965251
	lexp						
	L1.	-0.5734921	0.3087667	-1.86	0.088	-1.246237	0.0992527
	L2.	0.4846289	0.3041111	1.59	0.137	-0.1779723	1.14723
	L3.	-0.6335936	0.3799602	-1.67	0.121	-1.461456	0.1942685
	L4.	0.4779192	0.286494	1.67	0.121	-0.1462978	1.102136
	limp						
	L1.	1.114489	0.3414527	3.26	0.007	0.3705273	1.85845
	L2.	0.2744973	0.3286085	0.84	0.420	-0.4414791	0.9904737
	L3.	-0.1135789	0.3781564	-0.30	0.769	-0.9375111	0.7103532
	L4.	-0.6424163	0.2775102	-2.31	0.039	-1.247059	-0.0377735
	_cons	29.99879	7.786443	3.85	0.002	13.03359	46.964

(continued)

(continued)

lexp							
	lgdp						
	L1.	-2.997558	1.603116	-1.87	0.086	-6.490447	0.4953306
	L2.	5.823606	2.116247	2.75	0.018	1.212701	10.43451
	L3.	-0.1938903	2.002193	-0.10	0.924	-4.556294	4.168513
	L4.	-3.28821	1.466621	-2.24	0.045	-6.483702	-0.0927179
	laid						
	L1.	0.0634793	0.1079112	0.59	0.567	-0.171639	0.2985977
	L2.	-0.1575643	0.1402465	-1.12	0.283	-0.4631352	0.1480066
	L3.	0.1875782	0.1135864	1.65	0.125	-0.0599052	0.4350616
	L4.	0.0101117	0.1133523	0.09	0.930	-0.2368618	0.2570852
	linv						
	L1.	-0.0487891	0.125531	-0.39	0.704	-0.3222976	0.2247195
	L2.	0.5588016	0.0997545	5.60	0.000	0.3414551	0.776148
	L3.	-0.0260649	0.115656	-0.23	0.825	-0.2780576	0.2259279
	L4.	-0.4682789	0.1289864	-3.63	0.003	-0.7493161	-0.1872417
	lexp						
	L1.	0.9408638	0.2834283	3.32	0.006	0.3233267	1.558401
	L2.	-0.8131367	0.2791547	-2.91	0.013	-1.421363	-0.2049108
	L3.	0.54837	0.3487794	1.57	0.142	-0.211555	1.308295
	L4.	-0.5450935	0.2629834	-2.07	0.060	-1.118085	0.0278981
	limp						
	L1.	-0.3758638	0.313432	-1.20	0.254	-1.058773	0.3070458
	L2.	1.217207	0.3016418	4.04	0.002	0.5599864	1.874428
	L3.	-0.5559258	0.3471237	-1.60	0.135	-1.312243	0.2003917
	L4.	0.5965138	0.2547368	2.34	0.037	0.04149	1.151538
	_cons	6.711469	7.147462	0.94	0.366	-8.861512	22.28445
limp							
	lgdp						
	L1.	-0.8445355	1.607258	-0.53	0.609	-4.346449	2.657378
	L2.	2.560231	2.121715	1.21	0.251	-2.062588	7.183051
	L3.	2.632783	2.007366	1.31	0.214	-1.740892	7.006459
	L4.	-3.375885	1.47041	-2.30	0.040	-6.579633	-0.1721363
	laid						
	L1.	-0.1803269	0.10819	-1.67	0.121	-0.4160528	0.0553989
	L2.	0.0303954	0.1406089	0.22	0.832	-0.2759651	0.3367558
	L3.	-0.0405265	0.1138798	-0.36	0.728	-0.2886494	0.2075963
	L4.	0.0363696	0.1136452	0.32	0.754	-0.2112421	0.2839812
	linv						
	L1.	0.1705244	0.1258553	1.35	0.200	-0.1036909	0.4447396
	L2.	0.3524406	0.1000123	3.52	0.004	0.1345326	0.5703486
	L3.	0.0864899	0.1159548	0.75	0.470	-0.166154	0.3391338
	L4.	-0.1157233	0.1293197	-0.89	0.388	-0.3974867	0.16604
	lexp						
	L1.	0.1346757	0.2841606	0.47	0.644	-0.484457	0.7538084
	L2.	0.1782248	0.279876	0.64	0.536	-0.4315726	0.7880222
	L3.	0.140658	0.3496805	0.40	0.695	-0.6212305	0.9025464
	L4.	-0.3259262	0.2636629	-1.24	0.240	-0.9003983	0.2485458
	limp						
	L1.	0.1303626	0.3142418	0.41	0.686	-0.5543115	0.8150367
	L2.	0.1527984	0.3024212	0.51	0.623	-0.5061207	0.8117175
	L3.	-0.1198645	0.3480206	-0.34	0.736	-0.8781362	0.6384071
	L4.	0.4668792	0.255395	1.83	0.092	-0.0895786	1.023337
	_cons	-8.961236	7.165929	-1.25	0.235	-24.57445	6.651982

References

Abah, D., & Naakiel, P. W. (2016). Structural Adjustment Programme in Nigeria and its Implications on Socio-Economic Development, 1980–1995. *The Calabar Historical Journal, 6*(2), 1–7.

Bain, K. A, Booth, D., & Wild, L. (2016, September). *Doing Development Differently at the World Bank: Updating the Plumbing to Fit the Architecture.* Overseas Development Initiative. https://www.odi.org/sites/odi.org.uk/files/resourcedocuments/10867.pdf.

Bakare, A., Tunde, A., & Bashorun, O. (2014). The Two Gap Model and the Nigerian Economy; Bridging the Gaps with Foreign Direct Investment. *International Journal of Humanities and Social Science Invention., 3*(3), 01–14.

Burnside, C., & Dollar, D. (2000). Aid, Policies, and Growth. *American Economic Review, 90*(4), 847.

Chenery, H. B., & Strout, A. M. (1966). Foreign Assistance and Economic Development. *American Economic Review, 56*(4).

Chukwuemeka, E., Okechuku, E., & Okafor, U. (2014). Foreign Aid to Nigeria and Domestic Obstacles: A Review of Anambra State Education Sector. *Africa's Public Service Delivery and Performance Review, 2*(1), 52–81.

Delessa, K. (2012). *The Impact of Foreign Aid on Economic Growth of Ethiopia.* M.Sc. Dissertation, Jimma University School of Graduate Studies.

Dollar, D., & Easterly, W. (1999). The Search for the Key: Aid Investment and Policies in Africa, World Bank Policy Research Working Paper. No. 2070.

Fasanya, I. O., & Onakoya, A. B. (2012). Does Foreign Aid Accelerate Economic Growth? An Empirical Analysis for Nigeria. *International Journal of Economics and Financial Issues, 2*(4), 423–431.

Ferraro, V. (2008). Dependency Theory: An Introduction. In: Giorgio Secondi (Ed.), *The Development Economics Reader* (pp. 58–64). Routledge. https:/data.oecd.org/oda/net-oda.htm

Kalu, K. (2018). *Foreign Aid and the Future of Africa.* Palgrave Macmillan.

Kilman, J., & Lundin, J. (2014). Testing the Big Push Hypothesis-The Case of Montserrat. M.Sc Dissertation, Lund University school of Economics and Management.

Kolawole, B. O. (2013). Foreign Assistance and Economic Growth in Nigeria: The Two-Gap Model Framework. *American International Journal of Contemporary Research, 3*(10).

Liew, C. Y., Mohamed, M. R., & Mzee, S. S. (2012). The Impact of Foreign Aid on Economic Growth of East African Countries. *Journal of Economics and Sustainable Development, 3*(12), 129–138.

Mbah, S., & Amassoma, D. (2014). The Linkage between Foreign Aid and Economic Growth in Nigeria. *International Journal of Economic Practices and Theories, 4*(6), 1007–1017.

Mustafa, M. E., Elshakh, M. A., & Ebaidalla, E. M. (2018). Does Foreign Aid Promote Economic Growth in Sudan? Evidence from ARDL Bounds Testing Analysis. In: *A paper Prepared for ERF' 24th Annual Conference, 2018 The theme of International Economics* (pp. 1–25).

Nwosu, C. U. (2018). *Nigeria's Economic Growth: Does Foreign Aid Really Matter?* M.Sc Dissertation, The University of Sheffield.

Okoro, C. O., Nzotta, S. N., & Alajekwu, U. B. (2019). Effect of International Capital Inflows on Economic Growth of Nigeria. *International Journal of Science and Management Studies (IJSMS), 2*(1), 13–25.

Okpanachi, C. C. (2011). *Impact of Foreign Aid on Savings in Nigeria*. M.Sc Dissertation, University of Nigeria, Nsukka.

Olanrele, I. A., & Ibrahim, T. M. (2015). Does Developmental Aid Impact or Impede on Growth: Evidence from Nigeria. *International Journal of Economics and Financial Issues, 5*(1), 288–296.

Panjak, A. K. (2005). Revisiting Foreign Aid Theories. *International Studies, 42*(2), 103–121.

Siddique, H. M. A., Kiani, A. K., & Batool, S. (2017). The Impact of Foreign Aid on Economic Growth: Evidence from a Panel of Selected Countries. *International Journal of Economics and Empirical Research, 5*(1), 34–37.

Tang, K. B., & Bundhoo, D. (2017). Foreign Aid and Economic Growth in Developing Countries: Evidence from Sub-Saharan Africa. *Theoretical Economics Letters, 7*, 1473–1491. https://doi.org/10.4236/tel.2017.75099

Taylor, L. (1994). Gap Models. *Journal of Development Economics, 45*(1), 17–34.

Lawal, T., & Oluwatoyin, A. (2011). National Development in Nigeria: Issues, Challenges and Prospects. *Journal of Public Administration and Policy Research, 3*(9), 237–241.

Ukpong, U. A. (2017). Foreign Aid and African Development: Lessons from Nigeria. *Journal of Political Science and Public Affairs., 5*(3), 1–3.

Umoru, D., & Onimawo, J. A. (2018). National Policy and Big-Push Theory of Development in Nigeria: Moving Away from Low-Level Economic Equilibrium. *Scientific Papers of Silesian University of Technology, 116*(1995), 178–187.

Walin, A. (2014). *Aid, Policies and Growth in Sub-Saharan Africa: A Panel Data Study on Aid Effectiveness*. B.Sc. Dissertation, LUND University School of Economics and Management.

World Trade Organization (WTO). (2008). *Trade Policy Review Nigeria*. WTO.

Yiew, T. H., & Lau, E. (2018). Does Foreign Aid Contribute to or Impeded Economic Growth? *Journal of International Studies, 11*(3), 21–30.

CHAPTER 7

Gender Equality, Education and Mainstreaming of Gender in Ghana

Geeta Sinha and Bhabani Shankar Nayak

Introduction

'Deprivations in a wide range of economic, social and political dimensions reinforce the systematic exclusion of the poor and marginalised groups from rights to quality education, compounded by widening inequalities in many countries' (Subrahmanian, 2003: 1). The deprivation of women in the field of education is widespread. As a result, the 1990 Jomtein conference on 'Education for All' has witnessed actions by governments and development agencies to increase the educational enrolments with a special emphasis on girls. As a matter of fact, in the majority of developing countries, girls have lower enrolment rates, higher drop-out rates and lower achievement than boys do (UNESCO, 2003). The goal was further

G. Sinha (✉)
Oxford Brookes University, Oxford, UK
e-mail: gsinha@jgu.edu.in

B. S. Nayak
University for the Creative Arts, Epsom, UK
e-mail: bhabani.nayak@uca.ac.uk

B. S. Nayak (ed.), *Political Economy of Gender and Development in Africa*, https://doi.org/10.1007/978-3-031-18829-9_7

expanded and reaffirmed at Dakar in April 2000, to embrace issues of quality and achievement in the field of education. Again, the concern regarding girls' education was reiterated. The Millennium Development Goals further support the aspect of gender equality and women's empowerment. This discourse has led to new aid modalities. Budgetary support and SWAp are some initiatives to reduce gender inequalities in the field of education.

The chapter provides a brief overview of the gender perspective in the SWAp in the field of education. It also aims to explore SWAp and the extent to which they have the ability to integrate gender equality concerns in education. Generally, the implementation of SWAp varies extensively from country to country. Accordingly, the national governments are shouldered with the onus to manage and sustain their own development process by drawing support from international donor agencies. This approach has been strategically adopted by various agencies as an effective and efficient tool to ensure that development initiatives have national reach and ownership. This approach has been extensively used in mainstreaming gender in different fields. The development planners concerned with achieving gender equality in education are of the opinion that gender mainstreaming has been, and continues to be, successfully implemented within a SWAp.

The chapter analyses some of the ways in which mainstreaming gender has taken place through SWAp in the field of education in Ghana. This chapter is based on secondary data and an extensive literature review on mainstreaming gender and Sector-wide Approaches in the field of education. The chapter is divided into four parts to focus our study and put forth the arguments in a systematic way. The first section reveals the established importance of education and the existing gender inequalities in this sector. The sub-section of the first section also discusses the transition from Women in Development (WID) to Gender and Development (GAD), which is an important developmental discourse to analyse an issue. This section then leads to the second part of the essay, which deals with the concept of gender mainstreaming and its importance in the sector of education. The third part introduces SWAp as a strategy to address the existing disparities and inequalities while its sub-section highlights some of its efforts in the mainstreaming gender in the field of education. The final section of the chapter analyses the extent to which SWAp as a strategy have been able to mainstream gender in education in Ghana.

Education, Gender and Development

It is widely acknowledged that education plays a pivotal role in sustainable social and economic development. The field of education is always demarcated as a priority area of attention and investment, irrespective of the ideological orientation underlying approaches to development. According to Bellew and King (1993: 285), 'the benefits of education are by now well established. Education improves the quality of life. It promotes health, expands access to paid employment, increases productivity in market and non-market work, and facilitates social and political participation.' These benefits need to be experienced by both men and women in a fair and equitable way. They also ensure that women receiving education makes sense in terms of sustainable development and poverty reduction. Thus, education enhances the potential of women for contributing to the social, economic and political aspects of national development. This process has the ability to redress imbalances between women and men as well as other social groups.

Gender, in contrast to sex, refers to the non-biological aspects of men and women in the society. The 1995 Commonwealth Plan of Action on 'Gender and Development' defines gender as the socially constructed differences between women and men that result in women's subordination and in opportunity to have a better life (Commonwealth Secretariat, 1995). Women and men play different roles in the society through their relations to each other in the socioeconomic, political and cultural contexts. As per their different roles in society, they are treated differently as well. Such gender inequalities refer to the unequal relations between men and women, particularly in terms of access to power, opportunities and resources, be it at the household or at the societal level. So, it becomes evident that gender inequality exists at all levels in the society. Such inequality seems to be closely linked with poverty and there is no denial of the fact that the majority of the world's poor are women. In the process of eradicating poverty, the development planners have aimed to address the issue of gender inequalities. However, considerable gender inequalities exist in the education sector. Gender equality in education is one of the central poverty reduction strategies in the modern development discourse. Failure to provide education to girls is seen as a hindrance to anti-poverty measures and denial of human rights (Global Campaign for Education, 2003). It is also important to understand the major transition of GAD from WID in any developmental discourse to understand the approaches applied to address this issue.

From Women in Development (WID) to Gender and Development (GAD)

There has been a major shift from WID to GAD in the developmental discourse in regards to the issue of women, which is often discussed. By the late 1970s, the WID approach in the developmental discourse has been questioned because of its limited benefits to women and focusing on women's issues in isolation. Women were considered as subordinates and the most neglected and marginalized sections of the society. The WID approach tried to address this issue by implementing policies in programmes to 'include' women in the developmental process. Lack of access to resources and their unrecognized roles in production were considered as the key to their subordination. So, the programmes were designed around income generating schemes, micro credit and loans in order to increase their productivity. The analysis of women's subordination was the core issue of WID; however, the relational nature of subordination was largely unexplored. The WID approach viewed women more as 'recipients' in the developmental programmes rather than as 'active' agents of change. Women have become the 'pack horses' of development work, often working unpaid on development projects which exacerbated their triple burden of reproductive work, productive work and community managing (Moser, 1993).

In the 1980s, an alternative to the WID approach was introduced: the Gender and Development (GAD) approach. This approach tried to view the problems of women in terms of their gender (the social relationship between men and women in which women have been subordinated and oppressed) rather than in terms of their sex (their biological distinctions from men). It challenged the 'naturalized' roles of women in society and the dynamics of the household (Kabeer & Subrahmanian, 1999). So, gender roles could be understood by, 'examining and analysing the nature, basis and reproduction of male power' (Colclough et al., 2003: 6). The GAD approach supports the WID viewpoint of women being given equal opportunities to participate as men in all aspects of life, but its major focus is to examine the gender relations of power at all levels, so that interventions can bring about equality and equity between women and men in all spheres of life. The GAD approach also emphasizes legal reform (Commonwealth Secretariat, 1999: 12).

The WID approach has acknowledged the peripheral role in development and seeks to include women in the process. It tends to maintain the existing status quo of women in society. Alternatively, the GAD approach is more transformative in nature and attempts to challenge the status quo of the society and institutions that produce gender inequalities. Thus, the GAD approach does not merely integrate women into ongoing developmental initiatives, rather it seeks to bring about structural change and shifts in power relations in order to eliminate gender biases at all levels by mainstreaming gender. In the next section, we will be examining these approaches in reference to gender mainstreaming.

Why Gender Mainstreaming?

Gender mainstreaming can be defined as "the process of assessing the implications for women and men of any planned action, including legislation, policies or programmes, in all areas and at all levels. It is a strategy of making women's as well as men's concerns and experiences an integral dimension of the design, implementation, monitoring and evaluation of policies and programmes in all political, economic and societal spheres so that women and men benefit equally and inequality is not perpetuated" (United Nations, 1997: 28). 'Gender mainstreaming' as a strategy has been widely used in the development sector to overcome the problems related to exclusion of women from decision making at the policy level. It reflects the transition of broader literature about women and development from the Women in Development (WID) to the Gender and Development (GAD) approach. The increasing attention on the status of women in development can be attributed to the persistent activities of the women's movements throughout the 1980s and 1990s and the constant struggle to put women firmly on the development and human rights agenda (Stienstra, 1994). In the year 1979, the Convention on the Elimination of all forms of Discrimination Against Women (CEDAW) was signed. This convention has recognized the special human rights of women and holds all signatories/state parties liable for the status of women in their countries.

Furthermore, the series of international conferences organized by the United Nations are the results of the mobilization of feminist groups around the globe, irrespective of developed and developing countries. Such international conferences are the sources for the identification of strategies and policies to improve the conditions of the women in a wide range of social, political economic and cultural dimensions (Stromquist, 1997). In the Beijing Platform for Action, the term 'gender mainstreaming' was formally used for the first time and was recognized as the global strategy. It stated that actors' 'should promote an active and visible policy of mainstreaming a gender perspective in all policies and programmes so that, before decisions are taken, an analysis is made of all effects on women and men, respectively' (Beijing Platform for Action, 1995: para. 202). Integration of gender was to apply to all the projects, programmes and policies within the development framework. However, mainstreaming not only focused on the policies and programmes dealing with the effects on men and women but also led to the 'institutionalization' of gender. 'Mainstreaming looks beyond the promotion of projects and programmes for women, to the consideration of gender issues across all sectors, ministries and departments' (Byrne et al., 2002: 3).

Henceforth, mainstreaming not only refers to the policy level engendering process through programmes and projects but also triggers the transformation of governments, institutions and development agencies to be gender sensitive bodies. Kabeer and Subrahmanian (1999) argue that institutions need to change in order to alter the power and gender inequalities embedded in the society. The ultimate goal is to achieve gender equality with the aim of transforming structures of inequality (United Nations, 1997: 28). In practice, the institutionalization of gender within international development agencies and

governments has been varied and limited (Goetz, 1995). However, 'the persistence of gendered outcomes of everyday decision making in both state and non governmental organisations, often in spite of policy rhetoric promoting gender sensitive changes, has led to the conclusion that institutions themselves are gendered' (Goetz, 1997: 10). According to the Development Assistance Committee (DAC) 1999 guidelines, a mainstream strategy has two major aspects, namely: (i) the integration of gender equality concerns into the analyses and formulation of all policies, programmes and projects, (ii) initiatives to enable women as well as men to formulate and express their views and participate in decision making across all development issues. (DAC, OECD, 1999: 7) Thus, gender mainstreaming strategy includes initiatives specifically directed towards women. In the same way, initiatives targeted directly to men are necessary and complementary as long as they promote gender equality (DAC, OECD, 1999).

Gender Mainstreaming in Education

It is a fact that the world has missed the first and the most critical of all the Millennium Development Goals—gender parity in education: 'Without achieving gender equality for girls in education, the world has no chance of achieving many of the ambitious health, social and development targets it has set for itself' (Kofi Annan, March 2005 as cited in Global Campaign for Education, 2005). An estimated 60 million girls in the world are denied access to education. Investing in education is seen as one of the fundamental ways in which nation states and citizens can move together to achieve long-term development goals and improve both economic and social standards of living (Subrahmanian, 2002). This is borne out of data, which indicate that high levels of education and development are positively correlated (Schultz, 1994).

By achieving gender parity in education, countries could raise per capita economic growth by about 0.3 percentage points per year, or 3 per cent in the next decade (Psacharapoulos and Patrinos, 2002 as cited in Global Campaign for Education, 2005). Apart from economic growth, girls' education can contribute immensely to the reduction of hunger and malnutrition. Educated women are better able to resist practices such as female genital cutting, early marriage and domestic abuse by male partners (Herz & Sperling, 2004). This in turn helps them to have a better quality of life for themselves and their future generations. Through enhancing women's exposure to the outside of the four walls of the household, education is seen to have an implicit transformative effect, along with the opportunities it opens up in terms of employment (Dreze & Sen, 1995).

The education of women helps in securing intergenerational transfer of knowledge and providing ways to acquire long-term gender equality and social change (Subrahmanian, 2002). Henceforth, it seems imperative to have gender equality in education. Gender equality in education is seen as gender mainstreaming as the process of attaining equal access to schooling by the girls and boys, especially at primary levels and achieving equal and high levels of quality schooling. Equal access to schooling as such provides the necessary infrastructure to deal with the practical needs of women. It helps to move from a gender-neutral to gender sensitive approach to education. Although equal access to schooling is the key source for gender equality, it has failed to go far enough in addressing the broader issues of gender discrimination, both within and outside education. This discrimination contributes to subordinate the position of women in society.

Education is not sufficient in itself; it can easily overcome different forms of women's deprivation and oppression at many levels. The gaps and constraints in education in general, and female education in particular, are embedded in the social and economic structures and norms that obtain in different contexts (Fine & Rose, 2001). So, even educated women face embedded disadvantages in labour markets, property ownership and sexual and reproductive choices. However, education helps women to build their confidence amid persistent discrimination. Education also helps girls to understand about their rights and enhances their ability to acquire and process information and increased earning power (Global Campaign for Education, 2005).

Sector-wide Approaches: A New Strategy to Address Disparities and Inequalities Sector-wide Approaches (SWAp) were accepted as a new strategy to address the existing disparities and inequalities in different developing countries. They were first introduced in the early 1990s in response to the former developmental aid approaches, which resulted in not so satisfactory strategy in poverty eradication. This triggered the development planners to make a shift from project-based development to that of sector-based and multi-donor programmes. This transition reflexes the change in development thinking, from technical assistance and basic needs strategy to the theories of empowerment, which emphasize countries' ownership and accountability of programmes. Supporting individual projects were criticized which led to the fragmentation of the poverty reduction strategies as different donor/aid agencies had different priorities and agendas. The sustainability of such interventions was always questionable. Differing priorities between donor agencies and lack of communication led to poor co-ordination and reduced aid efficiency (Ratcliffe & Macrae, 1999).

The SWAp can be defined as "a process in which funding for the sector—whether internal or from donors—supports a single policy and expenditure programme, under government leadership, and adopting common approaches across the sector. It is generally accompanied by efforts to strengthen government procedures for disbursement and accountability. A SWAp should ideally involve broad stakeholder consultation in the design of a coherent sector programme … and strong coordination among donors and between donors and government" (ODI, 2001: 1). It seems apparent that through SWAp the donors work together in partnership with aid recipients to develop a 'coherent sector policy defined by an appropriate institutional structure and national financing programme' (Cassels, 1997: 11). It is both measurable and achievable. This in turn helps the donors to pool their resources together and work together for a common goal that includes proper financial planning and investment. Thus, SWAp is considered by the development planners to be a new and innovative strategy that promotes a sector development approach involving all the concerned actors to achieve a common national developmental goal. According to Ratcliffe and Macrae (1999), SWAp is "a new way of working together by improving working relationships, enhancing efficiency of development assistance, and improving relations with national authorities. It is also a new framework for development assistance including agreed action plans rather than separate agendas and contributes to the basket rather than fund separate projects. Moreover, it is a new instrument of development assistance which promotes sector reforms through specific, commonly agreed operational commitments, and provides greater authority to national governments in resource decisions (ibid: 42).

SWAp in the Field of Education towards Gender Mainstreaming

The emergence of sector-wide development processes as a shift in development thinking cannot automatically lead to an orientation of SWAp towards gender equality. One of the promising elements of SWAp is that it provides base and wider scope for equal rights and opportunities for both women and men. It follows an integrated approach rather than isolated efforts through projects. It involves shaping an entire sector with the objective of enhancing long-term development. Gender equality is one of the crucial aspects to be addressed by SWAp to be successful in meeting the goal of equitable and sustainable development.

However, the elimination of gender disparities in primary and secondary education by 2005 and achieving gender equality in education by 2015 are two international goals that draw attention for action as a part of donor partnerships in the field of gender and education. These goals have developed from the 1990 Jomtein World Conference on Education for All (EFA), and expanded in the follow-up World Education Forum (WEF) held in Dakar in 2000. The Millennium Development Goal (MDG) supports them for Gender Equality and Women's Empowerment.

Gender mainstreaming as a process needs to pervade the whole sector programme. It can provide a wider connotation to gender issues beyond increase in women's and girls' access to services and resources. Only through this wider approach of gender mainstreaming, can it recognize the structures of power relations, which impede the realization of gender equality (both within and outside development institutions) (Bell, 2000). It is challenging in the case of SWAp as an approach of gender mainstreaming. It is necessary to have a twin-track gender approach to understand this challenge and at the same time focus both on girls' education as a political priority and addressing the issue of male and female identities together with male and female education (Rose & Subrahmanian, 2005).

It is argued within a Sector-wide Approach to development such as SWAp, that effective gender mainstreaming is possible (Sibbons et al., 2000). SWAp provides an opportunity to address institutional bias through social analysis, which allows development professionals to understand how poor people are excluded from access to institutions within a particular sector (Norton & Bird, 1998).

Mainstreaming Gender through Sector-Wide Approaches in Education in Ghana

The many ways in which mainstreaming gender is incorporated in Ghana through SWAp in the field of education can be studied through the process of categorization. The process of categorization can be divided into three stages, namely, adoption of terminology, putting a policy into place and implementation (Moser & Moser, 2005). By analysing within the above framework, it can help us to understand the strengths and weaknesses of SWAp towards mainstreaming gender in the field of education in Ghana.

Adoption of the Terminology: Mainstreaming Gender through SWAp

The concept of 'gender mainstreaming' is being incorporated in a country-wide governmental programme in Ghana. The various studies and reports indicate that the DAC definition is used in which mainstreaming is where gender is considered in each and every activity of the government. Donors as well as government need to broaden their focus from access and participation to all the objectives of the sector programme. But in the education sector, the Education Strategic Plan (ESP) 2003–15 of the Ghanaian government adopts an issue-based approach to sector development by identifying four focus areas: (1) equitable access to education; (2) quality education; (3) educational management; and (4) science, technology and Technical and Vocational Education and Training (TVET) (Government of Ghana, 2003: 12). The Ministry of Education, Government of Ghana reiterates its commitment to a whole sector, or Sector-wide Approach to education development with a particular emphasis on girls' education. The ministry considers it a holistic approach to sector development, a process that includes the sector, the stakeholders and the beneficiaries in their entirety irrespective of gender (Government of Ghana, 2003: 33). This reflects the ethos of 'gender mainstreaming' through SWAp in the field of education.

The conceptualization and adoption of the terminology of 'gender mainstreaming' through SWAp in the field of education in Ghana, got its inspiration from the discourses of the Jomtein Conference (1990), conferences on 'Education for All', UNESCO, Dakar (2000), MDGs and the major policy advocacy agencies like DFID, OECD and the Commonwealth Secretariat (1999) have also inspired the process.

Contextualizing Gender Mainstreaming through SWAp in Ghanaian Education Policy

The Ghanaian education policy goal reveals that it aims to increase female participation in the education sector, in terms of enrolments, retention and completion rates. It also takes up the sensitization programmes highlighting the importance of female education. In Ghana, one-fifth of eligible primary school children do not enter school (over two-fifths in the northern, upper east and upper west regions) and many more young people, particularly girls, drop out of school without formal certificates or achieving functional literacy. Gross Enrolment Ratios in junior and senior secondary schools, at 64 per cent and 18 per cent respectively, are unsatisfactory. This condition of education in Ghana is a matter of national concern. To overcome this situation, and therefore to improve educational attainment, there needs to be serious action at every level. Access to primary and junior secondary education needs to be expanded. The growth in the basic education sub-sector can put more pressure on the secondary and tertiary sub-sectors like vocational/technical and science education. This has led to the targeting of TVET as a sub-sector. Thus, there is a long-term need and consensus to provide for greater access to secondary and tertiary institutions in a cost-efficient way. The policy also reaffirms its objective on quality education and better educational management for the improvement of the educational condition of the country. However, the above educational policy objectives and commitments reflect a sectoral approach for gender mainstreaming in education. But the study of the policy implementation and its outcomes can guide us to evaluate the effectiveness of SWAp in mainstreaming gender in education.

Implementation

In Ghana, the sector of education too faces gender differences in terms of access to education and number of male and female teachers. The females are the ones in the most disadvantageous position. This has triggered educational reform in 1987 towards the access of better quality education for all the children. Major focus was given to girls' access to Science, Technology and Mathematics as part of an explicit policy effort of the Ghanaian government to mainstream gender. Even the community was mobilized to bring in their participation so that 'bottom–up' planning and intervention of the programme would make it more people friendly and the Government would seem more committed.

Moreover, the establishment of the National Development Planning Commission (NDPC) in 1990 was an effort to bring about decentralized planning and enhance grassroots community participation. The process has led to the recruitment of District Girls' Education Officers (DGEOs) in each of the 110 districts of Ghana. Basic orientation training has been provided to all these officers, and a programme to provide training in Participatory Learning and Action (PLA) can gradually include all of them. The view was taken that problems related to girls' education were highly localized. Again, the introduction of free, Compulsory and Universal Basic Education (fCUBE) in the year 1995 was an effort to realize a Sector-wide Approach in education. Sibbons and Seel (2000: 18) reported fCUBE as 'a comprehensive sectorwide programme designed to provide good quality basic fCUBE education for all children of school-going age in Ghana by the year 2005'. SWAp is the subsectoral programme of fCUBE. Its objectives are to enhance access, quality and equity—although the fCUBE documentation does not outline a coherent strategy for achievement of these goals (Sibbons et al., 2000). This has led to the failure of the fCUBE in achieving the target. Another major attempt was the establishment of the Girls' Education Unit of Ghana Education Service

(GES). Following the 1994 UN mandate on the requirement of a girls' education focus within the Education For All (EFA) strategies, a Task Force on girls' education was established in Ghana. It aims to improve girls' participation and decrease drop-out levels, raises awareness about girls' educational matters at local and national levels and maintains girls' regular attendance in schools. Thus, it has created a positive and enabling environment for mainstreaming gender in education.

Government Education Strategy Plan documents include gender specific targets on participation, and objectives which are gender specific, but there are few textual references to gender and few if any activities beyond those directly related to access and participation (Sibbons & Seel, 2000).

Conclusion

Gender mainstreaming involves the processes of integration and transformation as their striking characteristics. The integration of women into development planning is a key element of gendered development. However to achieve gender equality, the involvement of institutional and organizational change becomes inseparable to gender mainstreaming. In this essay, efforts were made to explore the extent to which Sector-wide Approaches to education can foster successful integrative and transformative mainstreaming initiatives. This requires changes not only in material conditions of women, but also changes in the formal and social structures that maintain inequality. The institutions may also be transformed, so that gender equality is firmly placed on the agenda.

The efforts to mainstream gender through Sector-wide Approaches in education can be seen as a positive step in Ghana. Some of the educational policy documents reflect the ethos of mainstreaming gender through SWAp in the field of education in Ghana but often criticized as inconsistent policy formulation and implementation. The implementation of SWAp in education is a part of the policy commitments to gender mainstreaming in policy planning and implementation process in Ghana but is confined within few activities rather than incorporating the whole process. The emphasis continues to be on the concerns with females rather than concern with gender. There is little gender analysis done in the form of SWOT1 associated with the planning and implementation of activities that is one of the major drawbacks in the SWAp to education in Ghana.

Evidence from the Ghanaian Education Sector-wide Programmes illustrates that participation and access to education has been made possible but the transformation of institutions at the higher level is still grim. The introduction of the Girls' Education Unit and its responsibilities is unique and innovative in itself and can strengthen the gender mainstreaming efforts.

References

Beijing Platform for Action. (1995, September 15). *Fourth World Conference on Women: Action for Equality, Development and Peace*. Beijing.

Bell, E. (2000). *Emerging Issues in Gender and Development: An Overview*. Bridge Development – Gender, IDS.

Bellew, R. T., & King, E. M. (1993). Educating Women: Lessons from Experience. In E. M. King & M. Hill (Eds.), *Women's Education in Developing Countries: Barriers, Benefits and Policies*. Johns Hopkins University Press, published for World Bank.

Byrne, B., Bell, E., Laier, J. K., Baden, S., & Marcus, R. (2002). *National Machineries for Women in Development: Experiences, Lessons and Strategies for Institutionalizing Gender in Development, Policy and Planning. Bridge Report o. 36*. IDS.

Cassels, A. (1997). *A Guide to Sector-wide Approaches to Health and Development: Concepts, Issues and Working Arrangements*. WHO, DANIDA, DFID & EC.

Colclough, C., Al-Samarrai, S., Rose, P., & Tembon, M. (2003). *Achieving School for All in Africa: Cost, Commitment and Gender*. Ashgate.

Commonwealth Secretariat. (1995). *The 1995 Commonwealth Plan of Action on Gender and Development: A Commonwealth Vision Agreed in Principle*. Commonwealth Secretariat.

Commonwealth Secretariat. (1999). *A Quick Guide to Gender Mainstreaming in Education. Gender Management System Series*. Commonwealth Secretariat.

DAC, OECD. (1999). *Gender Equality in Gender Wide Approaches: A Reference Guide*. OECD.

Dreze, J., & Sen, A. (1995). *India: Economic Development and Selected Regional Perspectives*. Oxford University Press.

Fine, B., & Rose, P. (2001). Education and the Post Washington Consensus. In B. Fine et al. (Eds.), *Development Policy in the Twenty-first Century: Beyond the Post Washington Consensus* (pp. 155–181). Routledge.

Global Campaign for Education. (2003). *A Fair Chance: Attaining Gender Equality in Basic Education by 2005*. Johannesburg.

Global Campaign for Education. (2005). Girls Can't Wait: Why Girls. *Education Matters and How to Make It Happen Now', Reproductive Health Matters Journal, 13*(25), 1–8.

Goetz, A. M. (1995). *The Politics of Integrating Gender to State Development Processes: Trends, Opportunities and Constraints in Bangladesh, Chile, Jamaica, Mali, Morocco and Uganda. Occasional Paper 2.* UNRISD.

Goetz, A. M. (1997). Introduction: Getting Institutions Right for Women in Development. In A. M. Goetz (Ed.), *Getting Institutions Right for Women in Development, ch. 1.* Zed Books.

Government of Ghana. (2003). *Education Strategic Plan, 2003–2015. Volume 1: Policies, Targets and Strategies.* Ministry of Education.

Herz, B., & Sperling, G. (2004). *What Works in Girls' Education.* Council on Foreign Relations.

Kabeer, N., & Subrahmanian, R. (Eds.). (1999). *Institutions, Relations and Outcomes: A Framework and Case Studies for Gender-aware Planning.* Kali for Women.

Moser, C. (1993). *Gender Planning and Development: Theory, Practice and Training.* Routledge.

Moser, C., & Moser, A. (2005). Gender Mainstreaming since Beijing: A Review of Success and Limitations in International Institutions. *Gender and Development Journal, 13*(2), 11–22.

Norton, A., & Bird, B. (1998). *Social Development Issues in Sector Wide Approaches. Working Paper 1.* Social Development Division, DFID.

ODI. (2001). *Key Sheet for Sustainable Livelihoods. No. 7: Sector Wide Approaches.* ODI.

Ratcliffe, M., & Macrae, M. (1999). *Sector Wide Approaches to Education – A Strategic Analysis. Paper No. 32.* Education Research Series, DFID.

Rose, P., & Subrahmanian, R. (2005). *Evaluation of DFID Development Assistance: Gender Equality and Women's Empowerment. Phase II: Thematic Evaluation: Education. Working Paper 11.* DFID.

Schultz, T. P. (1994). *Human Capital Investment in Women and Men: Micro and Macro Evidence of Economic Returns. Occasional Paper 44.* International Centre for Economic Growth.

Sibbons, M., & Seel, A. (2000). *Mainstreaming Gender through Sector Wide Approaches in Education, Ghana Case Study.* Report for DFID, ODI, London and Cambridge Education Consultants.

Sibbons, M., Swamfield, D., Poulsen, H., Giggard, A., Norton, A., & Seel, A. (2000). *Mainstreaming Gender through Sector-wide Approaches in Education: A Synthesis Report.* Report for DFID, ODI, London & Cambridge Education Consultants.

Stienstra, D. (1994). *Women's Movements and International Organisations.* Macmillan Press.

Stromquist, N. P. (1997). Gender Sensitive Educational Strategies and Their Implementation. *International Journal of Educational Development, 17*(2), 205–214.

Subrahmanian, R. (2002). *Gender and Education: A Review of Issues for Social Policy, Social and Policy Development. Paper No. 9.* United Nations Research Institute for Social Development.

Subrahmanian, R. (2003). Introduction: Exploring Process of Marginalisation Inclusion in Education. *IDS Bulletin, 34*(1), 81–89.

UNESCO. (2003). *Gender and Education for All: The Leap to Equality. Summary Report. EFA Global Monitoring Report, 2003–04.* UNESCO.

United Nations. (1997). *Report of the Economic and Social Council for 1997. UN Document Symbol-A/52/3/Rev.1/Add.1.* UN.

CHAPTER 8

Sexual and Gender Based Violence in Displaced Contexts: Narratives of Somali Refugee Women and Girls in Dadaab, Kenya

Fathima Azmiya Badurdeen

Introduction

It is a well-accepted supposition that women are the major victims or survivors of Sexual and Gender based Violence (SGBV) in refugee camp settings, while acknowledging men and boys too are vulnerable to sexual violence. SGBV in refugee settings include acts such as rape, intimate partner violence (IPV), marital violence, early or forced marriages and sexual exploitation (Freedman, 2016; Simon-Butler & McSherry, 2018; Ozcurumez et al., 2020). These forms of violence are deeply embedded in specific cultural contexts wherein harm inflicted is supported directly or indirectly through structures and ideologies that permit specific forms of violence to continue in its precise contexts (Jensen, 2019; Sokoloff & Dupont, 2005). However, regardless of the universal nature of SGBV in

F. A. Badurdeen (✉)
Department of Social Sciences, Technical University of Mombasa,
Mombasa, Kenya
e-mail: fazmiya@tum.ac.ke

B. S. Nayak (ed.), *Political Economy of Gender and Development in Africa*, https://doi.org/10.1007/978-3-031-18829-9_8

refugee settings, it is also shaped by the values and circumstances of particular African cultures. Violations of women's rights are justified in the name of culture (Storkey, 2015; Wendt & Zannettino, 2014). Success in countering SGBV requires the integration of a gender and cultural lens to understand the violent act as well as to build locally relevant frameworks for intervention (Chynoweth, 2017; Cahn & Aolain, 2009; Badurdeen, 2020).

It is in these forced migration contexts, this chapter seeks answers for two interrelated questions: how does women's positioning in culture interpret SGBV in refugee camps? And, how does such positioning in culture explain women's resilience in the SGBV aftermath? At the core of these questions lay the common problem of culture between global discourses on SGBV and local cultural understanding of gender, power and violence as an extension of violence attributed to women that existed during pre-war or peaceful times. This chapter uses the case of Somali refugee women in Dadaab Refugee Complex to add to the scholarship of understanding the nexus of women's positioning in Somali culture and the SGBV context to investigate issues surrounding African women and culture through two central themes: sexual, gender based violence and resilience. The chapter begins by exploring the research approach on locating the study in the research design and framing the research questions. This is followed by an analysis on SGBV among Somali refugee women in Dadaab refugee camp examining the various forms of violence and impact of such acts through an analysis of women's positioning embedded in Somali culture. The second section of the analysis considers the challenges and prospects associated with Somali refugee women and their resilience as survivors in the aftermath of SGBV. Finally, the study looks at support systems for SGBV survivors enabling them to find a voice or their resilience to survive.

Locating the Context

This chapter is situated in the context of an emerging scholarship of SGBV in conflict and refugee settings. In an attempt to review studies on culture and violence related to Somali refugees I draw on the work of Giles and Hyndman (2004), and Cockburn (2004) whose studies posit that violence attributed to women does not take place only during conflicts but also in the absence of war. Cockburn (2004) argued that the sites of gendered violence are intertwined. This has been reiterated by Giles and

Hyndman (2004) who ascertains that violence on women in war does not exist isolated in a particular time or location but is reflected in a process which can be linked to gender relations in the households. The continuum of violence stretches from existence of violence in the everyday life, through the structural violence of socio-economic and political systems that sustain inequality for women which become further exacerbated during armed conflicts. The understanding of SGBV in refugee settings need to be understood through a continuum of violence which is intensified during forced migration contexts but have been in existence during peacetimes (pre-war context) (Cockburn, 2019; Menjívar & Perreira, 2019; Chakrabarti, 2017). In forced migration contexts, in spite of this continuum being part of a refugee woman's life, violence may also be experienced as a 'shattering experience of discontinuity' (Walker, 2009: 29) occurring within the inequalities on a daily basis. Within the core of this continuum lies the aspect of culture where the positioning of women is situated through structural (patriarchy) and cultural (through norms and religion) values (Galtung, 1971) where Kelly (1993) considers sexual violence as the most successful form of patriarchal control that constrains women's lives in public or private spheres. Krause (2015) acknowledges the need to recognize the continuum of violence across forced migration settings from pre- conflict to encampment and, during the stages of flight to elucidate the complexities and compounding risks faced by women in forced migration. Protection against SGBV span even after resettlement, repatriation or integration due to the continued prevailing structural and cultural violence for women and girls (Canning, 2017; Buscher, 2017).

A qualitative research approach within a narrative research design was used gathering, investigating and analysing stories of experiences of Somali women. A storytelling approach was used to collect narratives where the participants had the ability to communicate and to reinterpret their life experiences through stories (Reissman, 1993). The case centred life stories collected using in-depth conversational interviews facilitated a narrative inquiry to gather and examine stories on women's SGBV experiences to understand women's lived experiences of trauma, tolerance of violence, strength and resilience of victims and explore potential for providing a strategy for intervention through the understanding of the continuum of violence that existed during peaceful times (pre-war) and the forced migration contexts (refugee camps). The fieldwork resulted in an individual level study on women exposed to SGBV. This chapter does not intend to represent the views of individuals or try to capture the entire communities' perceptions of sexual and gender based violence and resilience. Rather, it captures narratives of few women who had willingly expressed their concern to share their stories. The aim was not to generalize findings, but to look at contextualized experiences. Participants were engaged using a purposive sampling combined with women community leaders as gatekeepers when possible. Twelve cases were selected for this study based on the criterion of saturation, where the sampling was discontinued when no new themes merged from the narratives. The chapter was also complemented through FGDs conducted during the study period. Pseudonyms are used in this chapter for participant anonymity.

Sexual and Gender Based Violence against Women in Dadaab

In the Dadaab refugee camps, Somali women have been subjected to multiple levels of discrimination and SGBV in both private and public spheres, wherein men took control of women's lives and resources (Abdi, 2006). These studies reveal the complex nature of SGBV in the Dadaab refugee camps ascertained as explained by two refugee women, through the following two narratives of Aamina and Yasmiin:

> *I still have those nightmares. I struggle to forget, but it is difficult. I was raped by armed men in Kismayu (Somalia) when I fled the conflict. I fled with my neighbours to Kenya. It was a very long and difficult journey. I reached and settled in the camp. My family members could not make it to the camp.*[1]
>
> *I was forced to marry at the age of fourteen. My parents forced me to marry as they felt a girl was a burden and needed to be married quickly for protection. Parents fear that they will not find a potential match while living in these camps where there are less potential for suitors. It is more cultural. My parents found a suitor who had conveyed his interest for me. He was in fifties and was previously married. Since I was given a good Mahr (dowry) my parents agreed. I was not happy with my marriage. My husband forces himself on me. I can't complain. It's not correct to complain in my culture, as we serve our husbands.*[2]

The narratives of Aamina and Yasmiin are subjective and selectively used to explain different forms of abuse women undergo in refugee migration settings. The two cases provide selective memories of the incident (sexual-based violence) unfolding the subjugated knowledge of the pain to explain the offence. For, the two narratives reveal two different contexts of sexual violence in refugee setting apart from their experience of being forced migrants the narratives capture the three forms of sexually and gender-based violence that will be explored in chapter.

SGBV in Dadaab refugee camps attributed to women is closely associated with the patriarchal structure visible in the Somali culture of refugees. Women are confined to a submissive role and are invisible in the public domain. Existing patriarchal and patrilineal traditions have confined them to the domestic sphere neglecting their participation in education, politics and economics. Apart from this structural violence, even direct violence against women remains a private affair restricted at a family level in the Somali culture. The physical abuses undergone by women within family homes were not considered as a violation of rights in the Somali communities. Majority of people believed that violence is rare and non-existent within the Somali family. This is well posited in the work of Islan (2015) that explain on the social organizational structure of Somali society is based on how men define their public domain. Within the framework of refugee settings, forced and early marriages take place for varied reasons such as poverty within the refugee families, the need for protection for the girls, girls exchanged for goods and money by the family, a girl handed over by the family to another to settle disputes within families or to preserve the honour of the family where the girl will have to marry the rapist.

Places of Violence

The narrative of Aamina revealed that the sexual violent act as an 'external act' committed by in the public domain where violence was committed outside her residential space. Yasmiin's narrative revealed sexual violence as an act within a consensual relationship (the private domain) in her own residential space. The dimension of 'space' for women is intricately entwined with her culturally allocated position of 'women being in the house working in her household chores' and her responsibilities 'placed on her like going out to collect firewood'. These spaces for women are places that need intervention in terms of protection. A FGD discussion highlighted the following as contributory factors for SGBV among women and girls: the lack of toilets and the need to use spaces away from the camps, during distribution of food or when trying to access their shelters, lack of security within and outside the camps, lack of proper locks in some shelters and the poor lighting systems in the night.[3]

Perpetrators and Victims

In the case of Aamina, the sexual violent act was committed by an external actor such as armed men and in the case of Yasmiin, it was the victim's husband and parents who married her off at an earlier age. The two cases show 'the need for protection' outside and within the camp—mainly in the private sphere. In the first case, the need for protection is from 'an external actor or outsider' while the second case highlights the need for protection from an insider who is the husband of the victim. Usually, the first case can highlight the plight and gain support from the community while the second case may not generate the same support. The community may not view marital rape or underage marriages as violence against women. The Somali culture purports women in a position wherein as the wife it is her inherent right to listen to her husband and has the ability to give into his desires such as sex or as a daughter, it is her duty to respect the father or the elders of the community. Hence, it was a duty to confirm to community obligations or the wishes of the father as in the case of early marriage. The two narratives also show the need for different intervention in the two contexts—in the public and private sphere.

In early and forced marriages, perpetrators can be women as well. Women in FGDs revealed that the earliest age for a girl to marry is 10–12 years or when the girls started their first menstruation. This was considered as per their culture and religion. Girls married early mainly because of parental or community pressure, peer pressure, poverty in the camps or the fear of being raped or sexually abused. The choice of the husband is decided by the parents. When parents choose their husbands, the girls are obliged to marry them even if they do not like them. The perpetrators in this case are the father or elders. In some cases, the mother too was blamed as they perpetuate this culture of early marriage depriving the girls of education or freedom of choice. Further, early and forced marriages often lead to other forms of sexual violence such as marital rape and other forms of intimate partner violence. Rape was seen as the most violent form of SGBV in refugee settings where they claimed rape in any form is unacceptable. Yet, marital rape was questionable as participants were unsure whether it was a crime as expressed during the FGDs.[4]

Impact on Gender Relations

The narratives highlight the following elements in the act: 'women's positioning, violence as control, male domination or a patriarchal structure, the power dynamics in the relationship and lack of consent and perpetrator/perpetrators being a man. The narratives also set the ground to understand the cultural context of women's positioning in Somali culture, where women's positioning in Somali culture, where women conceal their plight and in the case of Aamina, knowing the act is an abuse and the case of Yasmiin unknowing the act is an abuse considering it as 'normal' in their life.

The highlighted statements on the impact of violence, brings forth the following dimensions: (i) the patriarchal structure of the society and how it views SGBV. Here it was viewed that the man was dominant and had the strength and capability to do the act (rape/IPV) on women. (ii) feeling unsafe in the camp where women fear future attacks in the camp settings related to the remembrance of the previous attack (iii) the fear of women going out of the camp for collecting firewood due the fear of being raped or assaulted. Women may avoid situations with men for the fear of what some men are capable of doing (iv) for these women who were refugees being marginalized in the host country (especially among the surrounding host communities) sexual assault/rape reaffirms the assumption that they are devalued women 'insidious trauma' (Wasco, 2003: 311), and the outside world is unsafe and dangerous (v) religious and cultural interpretations on the husband-wife relationship was also evident highlighting that showed the very existence of women their positioning in the Somali culture of women subordination.

Effects of SGBV

The effects of these SGBV acts were physical, emotional and psychological. The immediate physical trauma revealed by the respondents included: injuries by being beaten, bruises, confusion, being shocked, having sleepless nights and being hyper-vigilant. In the case of Intimate Partner Violence (IPV), Nimo revealed that it 'happened all the time' and Nimo did also add that it was accompanied by beatings. Apart from the immediate physical trauma, the respondents highlighted certain psychological and emotional impacts during the interviews. Aamina and Ilhaan repeatedly said the statement 'I wish I was dead'. The memories of their attack constantly made them wish for death or the need to forget the incident or being paralysed by the fear of the encounter such as the case of Aamina who stated: 'he is a man, how can I fight him. Even if I tried, the other man will surely grab me'.[5]

FGDs revealed other responses such as fear of being pregnant or contracting any disease, the fear of being a refugee wherein her marginalized status makes her vulnerable to sexual assault or rape or being unsafe in the camp or outside the camp. According to Fowsio, a participant in her late thirties, felt that refugee women seemed more vulnerable when being outside the camp: 'we are refugees. They (host) can do anything to us. When we move out of the camps, we are vulnerable…I am sure he was not from the camp…I have not seen him before'. A participant from a FGD revealed that divorced and widowed are among the most vulnerable. 'all men tempt us (Divorcees) as partners for illegitimate relationships. We are considered as available and as sexual partners. In this community it is us (women) who would be blamed if people see us even talking to a man. Our interest is marriage. But who is interested in marrying a divorcee? …we have very little support. The lack of insecurity in the place makes us fear that we would be raped anytime – mainly in the night.'[6]

The aftermath of sexual violence can bring in a host of other problems associated to virginity and female genital mutilation (FGMs). Usually after rape, the woman is stitched up. This is done by elderly women in the community using traditional methods detrimental to the victim. The loss of virginity means she cannot marry a good partner where she loses all hopes to get married.[7]

Normalizing and Tolerating Violence

The religious and cultural influence of Somali society affects the way the refugee women victims normalize or tolerate the incident of SGBV. Two of my participants stated that the fear of being stigmatized and labelled as being raped or abused by the husband were reasons on why they never revealed the incident. While some kept the incident to themselves, other participants such as Idil revealed the incident to her immediate family. 'I told my mother and sister. Initially, they said that I was a disgrace to the family. I was not able to explain that it was not my fault that I was raped outside the camp'. The usual response when the issue is revealed to the family members is the view of the daughter, wife or sister bringing disgrace to the family. Community stigmatization is such that the family hides the girl or in extreme cases, it can result in the girl or woman committing suicide or even disappearing from the camp or village.[8]

In the case of normalizing and tolerating violence, self-blame was often cited as the outcome of the SGBV activity. The criteria of being a faithful woman, the honour of the family, self-sacrificing, a subordinate wife and mother surfaced as key cultural influences that depicted women in the society. As women are the honour of the families and communities, they remained in this abusive cycle of violence may it be early marriages or IPV. Rape victim is looked as if she is a curse, forbidden and not fit to be married again. Her tarnished image in the community can be double trauma for her—one on being raped and the other on how the community viewed her. According to Idil, community showers the blame of rape on the victim as it was her fault, 'rape is often blamed on the girl by the family or community. Some would go as far as saying that it is the girl's fault on provoking men'.[9]

Reporting/Under Reporting of Violence

Being stigmatized or being labelled prevented women from seeking or reporting the incident to the authorities. While some hid the issue, the others settled within the families or within their communities. Many preferred the Maslaha system compared to the Kenyan legal system. This was due to the lack of trust in the Kenyan Police (due to bribery and in some cases they are the perpetrators). The Maslaha system was a community-based institution to settle community disputes as well as rape and sexual assaults. Rape cases or sexual assaults were settled through the traditional courts headed by community elders who never understood the plight of refugee girls and women whose future have been destroyed. Elders come together and address the issue through the traditional norm (xeer). This has been the trend in the Somali community and is still reflected in the refugee setting. Here punishment for raping a virgin is more severe than a married woman or an old woman. The punishment is usually based on compensation for the victim's family which is based on the socio-economic status.[10] Blood compensation was given to the family of the victim (usually in the form of livestock or money) which never reached the girl, instead was handed to the male elders of the family, most commonly the father. Further, when a woman is raped, the 'problem' is settled by a meeting between traditional leaders and the woman's husband or family, and the perpetrator's family. 'They proceed to negotiate the monetary value of the "damage" done to the victim's husband or family's honour, where upon compensation is paid accordingly. If the victim is married, the rapist can sometimes be jailed if her husband refuses to accept compensation. In cases where the victim is unmarried, marriage of the victim by her rapist is the accepted solution. At no time is the victim consulted or even were present at these meetings' (UNIFEM, 2005).

This has severe disadvantages. The compensation varies from one context to another and it is often negotiated by the elders and the male family members. The compensation does not reach the victim but her family. There are reported cases that the Maslaha court is corrupt and has tendencies to favour parties based on benefits for the members of the system. Solutions by the Maslaha court as getting married to the perpetrator, as in the case of rape means, living with the man who had given a scar in her life. This, many women consider as unacceptable. Some opened up only when there was monetary gain or other material gains by organizations. Parents encourage their daughter's to reveal the issue if there was some assistance in the form of money or material gain from organizations.[11] Mwangi (2012: 22) highlights that, 'the community-based approach to resolving disputes based on compensation is a continuation of the subjugation and oppression of Somali refugee women.' These traditional courts are backed by a patriarchal structure, male dominance and have no gender equity. Culture always favoured the perpetrator in the traditional system of justice. Culture has implications on compensation in terms of blood money for the sexually violent act as women's positioning in society is undermined under customary law. Hence, a woman becomes vulnerable to the substitution of customary law to Shariah law. For example, a man cannot be killed for a woman since a man is worth 100 camels and a woman is only worth 50 camels. In case of injury, a woman's compensation is half of the man. Sexual crimes also go unreported as community elders exert pressure on women to settle through traditional channels rather than the courts forfeiting their legal rights.

Fear loomed as women believed that they will not be believed, or they are responsible for the attack. The resulting is the silence that surrounds such sexual violence bringing out a hidden prevailing culture of violence. In many ways, the rape victims are proved as being responsible for her condition. The raped or sexually assaulted victim may agonize herself 'being a woman' her behaviour that led to the attack forcing us to ask ourselves that, if these women censor their behaviour will it have avoided herself being targeted? Can this prevent such sexual assaults or rape? Is this not a case of double victimization where in spite of being a victim of violence the culture sheds all the blame on her? Some community members also do believe that false accusations are made on sexual assaults out of malice or fear of disapproval of consensual sex.

Women's Resilience in the Aftermath of SGBV

Resilience refers to an individual's ability to overcome psychologically, learn from and positively adapt to life's adverse events. In simple words, it means the ability to revert or bounce back to a point of equilibrium despite adversity (Lenette et al., 2013). Naturally, the resilience category has been well applied into the refugee communities as they succumb to major upheavals in rebuilding their lives. Pulvirenti and Mason (2011) critiqued the resilience concept, reasoning that refugee women were more than victims were, as well as more than survivors, which entailed that in the dynamic space of daily lives refugee women face a complex set of possibilities despite their victimhood status giving meaning to the process than the traits of resilience. There is much value in applying the resilience discourse to understand the experiences of refugee women as it affords an opportunity to move away from a victim status to a contribution that can be immense in understanding the SGBV experience of Somali refugee women in the Dadaab refugee camps. The narratives of these Somali women studied focus on four aspects: resilience situated in their ordinary daily lives, personal characteristics of the survivor, resilience promoted through informal and formal sources and social complexities of resilience and stress.

Resilience Situated in the Survivors' Ordinary Daily Lives

This 'daily life' of a victim is in itself an achievement of being a survivor amidst an individual-environment interaction. The women's sense of well-being stems from her interaction within the space between her and the environment. Within this interaction, it becomes futile to see the survivor's problem of the sexual violent act independently but locate it in a context of how the community views her problem, attaches cultural meaning to the act and making it a social problem. Surviving in this social problem is often not easy. Two strategies were used in terms of revealing the act for their survival on a day to day basis. Either they hid the act of sexual violence or they revealed it to their families or the communities. Hiding the issue and going on with their daily life was the most commonly used strategy among women where their family reputation was at stake like losing the love of their husband or not being able to find a suitor, if the victim was unmarried. Losing the love of the husband can be painful due to the violent act, as the women could not envision how their husbands would react to the incident. Sagal, a participant explained that it was painful, but better to hide the incident [rape] than bring disgrace to herself or her family. 'If you tell, you only expose yourself. Rather than assistance, relatives and community members will look down on you. We often hide it for our family honour'.[12]

Idil had the courage to open up to her mother and sister. The issue was out in the community and she became accustomed to the looks and community gossip after time. Inside within her, she knew she was not the one who wronged. She went about with her daily chores of cleaning, cooking and participating in community activities. Sometimes she cursed the perpetrator that he should be punished in hell, as her faith will not let her down. Finding solace in religious symbolism was a strategy undertaken by many of my participants. Aamina and Ilhaan included the statement 'God will give me a new life'. Faith gave them a new life or gave hope or expectations for a new life. There was a strong sense of faith, where women found their strength to endure. This included prayers where they felt as if they 'became closer to God.'[13]

Differing strategies employed by refugee women survivors of sexual violence hailed from denial of the issue or trying to forget the issue to speaking out and advocating for women's rights. Among the twelve, it was only Basra who wanted to fight for the rights of women, as she did not want other women to go through the same fate as encountered by her. This brings us to the next aspect of resilience—***the personal characteristics of the survivor***. Among the narratives, two catch words stood out under this aspect: 'courage or will to survive' and 'determination'. Four participants attached these words associated with their children and family life. They tried to forget the incident and move on with their life for a better tomorrow. According to Nimo, it was her children that gave her the courage to go on with her life. For Rehma, a participant who was in her mid-thirties, it was her determination to have a good family life with her husband that gave her the will to survive. Amidst their plight, these women were able to sacrifice their own happiness and need to uphold their cultural values and norms to redefine their situations as liveable, tolerable, reasonable, understandable and survivable.

Narratives of these refugee women survivors of sexual violence revealed that resilience was both supported through individual will and ***promoted through informal and formal support network and organizations.*** Resilience is both conditional upon the assistance of the community and constructed through support of community, family, friends and agencies (Chung et al., 2013). The participants revealed the importance of family support during the crisis. In cases where the family support was high, women found ways to cope and the ability to learn to accept themselves and to go on with their lives. Idil revealed that the support given by her mother and sister was important for her to move on with her life. Almost all of the respondents aimed for a new life. Family support was considered as the best support received, so that they did not feel isolated. Various counselling sessions carried out in organizations have helped the women in overcoming their plight and assisted them in the process of healing, reporting these crimes and seeking medical advice. Women who had some means of livelihood and had independence had the ability to cope with the situation and move on with their lives.[14]

Social Complexities of Resilience and Stress among Survivors

Narratives revealed that when the community was considered as a source of support for some women, the others felt it was a source of stress. The women experienced the support and stress in a highly gendered way. Basra revealed that she was associated with a community women's group that helped her and women of similar contexts to share their stories and find consolation among the group. The community group gave an opportunity for the women to discuss rights, break the silence of violence and the associated stigma. This group was supported by an NGO and was the basis for jealousy by other members of the community. Lack of acceptance of marital rape and early or forced marriages by some of the community members or associated stigmas on rape have affected the wellbeing of the women. Some women intended to isolate themselves from community activities, while others negotiated their lives within their cultural context of community shaping their world of being a woman.

Resilience-culture nexus shows that Somali women's experiences of SGBV and resilience strategies are culture-bound. It is important to understand the factors against which these women are resilient (e.g., sexism), where they build their day to day resilience to survive. Despite many women's experience of traumatic violence in displaced contexts, the analysis showed that displacement revealed individual and collective resilience in the aftermath of SGBV. Women aimed to live normal and meaningful lives in their everyday settings in camps. The motivation to live was particularly for their children and families.

Somali women's resilience strategies in the healing from the sexual violence act must explore the cultural context of gender associated to gender relations, women positioning in culture and traditional gender norms. Exploring how sexism has impacted resilience strategies in Somali refugee women is embedded in culture. For instance, Somali women may maintain not to disclose her sexual abuse due to cultural norms that demand women's silence in relationships as an attempt to maintain family relationships that can minimize discord between the private and public domains of her life in the camps. However, this silence may actually affect her psychologically, and affect her resilience strategies that are affecting her daily functioning in a negative manner. At the same time, surviving the trauma of sexual abuse and identifying the ways gender-role expectations have influenced her coping and survival styles is an area that needs to be explored.

The girls and women's narratives revealed how resilience was a process that operated in the social spaces that linked them to their environment where they embraced their own personal resources such as their inner will to survive and opportunities available outside such as NGO activities that supported them after the violence. Personal resources such as 'living for their children' and religious symbolism such as 'my faith kept me alive' seemed to dominate the resilience of many women. Informal social support roles from family members provided the necessary emotional and financial support for women. The emphasis on family in culture and its role as a support mechanism may encourage the development of strong interpersonal skills for resilience which is an important strategy for Somali women in their response to their sexual trauma in the SGBV aftermath.

Formal social support mechanisms, such as the forming of women groups—maintained and financed by NGO activities, assisted women with a platform to discuss their issues with like-minded individuals. The exploring an individual's resilience (the ability of the survivor to connect interpersonally with others) would allow for acknowledgement of "the complex, multilayered interpersonal and cultural dynamics that affect one's ability to be resilient". The survivors used their interpersonal skills to form supportive relationships outside the family, and within these safe relationships they were able to bring their plight and the associated shame of sexual abuse and place blame on the perpetrator.

Discussion

The merging themes on SGBV are looked through the context of Somalian society and its culture considering the intersections between religion, power and gender. While cultural explanations of SGBV are a contested terrain, cultural contexts are critical to the analysis of SGBV as they are always applicable as everyone who has a culture whether it was during peace times (pre-war) or in displaced contexts (refugee camps). Cultural explanations are contested because they are either used to excuse individual actions or used to facilitate stereotyping. Hence, culture is often responsible for how we view the problem of SGBV and how we address it (believing that women from a particular culture are passive, normalize and tolerate violence, fears to seek help, refrains to speak out about abuse, as it would shame the family and her resilience in the aftermath of the violence). Culture cannot be viewed in isolation from religion for there is an inextricable link between religion and culture.

The SGBV themes of how the violent act is viewed, tolerated, normalized, reported and compensated needs to be depicted through culture, religion and a patriarchal interpretation. These patriarchal interpretations are those of individuals in positions of power or means of maintaining power who define the dominant culture and interpretation of religion (Shaheed, 2009). It is important to view these themes through the power relations of the Somali society to shed light to the context of violence in refugee settings. For instance, marital rapes or early marriages in camps and reporting such acts to the authorities are dependent on women's positioning in Somali culture. In Dadaab refugee camps, behaviour is regulated under the Somali customary law (Xeer). Despite the Shariah (Islamic Law) coexisting with the Somali customary law there are contradictions between the two. In Somali culture there is no discrimination against women as the famous proverb goes on 'Gari Allay Taqaanna' meaning, 'people are equal in front of God'. In practice, customary law is pervasive and undermines the application of Islamic law.

Like many other cultures, Somalian culture is patriarchal and women's space within traditional settings is located within tribal and clan rules. While the fathers, brothers and male cousins make decisions in the domestic sphere, in the community decisions are made by councils of men. In this patriarchal society, women are only seen as a means for biological lineage and inter clan alliance (Islan, 2015). Prior to marriage, a woman is under the care and control of her father and after marriage, she comes under the control of her husband. Throughout her life, she is controlled by men and is protected by men and has no decision-making power (Garner & El-Bushra, 2004). This reveals powerlessness among women in Somali society. In such condition of powerlessness, marital rape is seen as an act within the private sphere and is not seen as a crime. Women's sexuality is seen as a property of the husband and sexual intercourse is not dependent on the consent of the wife; and within the institution of marriage, the husband has the exclusive right to their wives bodies. As reiterated by Grieff (2010) the institution of marriage is interpreted as a contract that provides financial maintenance of women in exchange for her obedience and sexual availability. While statistics are few in the camps, FGDs reveal marital rape as undeniably widespread and unregulated form of culturally justified violence on women.

Displaced contexts provided opportunities for changes as refugees are exposed to influences of international aid community and to ideas of gender equality and empowerment. Among the focused group discussions, it was interesting to find out that some of the members were able to highlight the forms of SGBV, their causes and consequences. For example, in the case of rape, they knew the ways of reporting to the authorities and seeking medical treatment. They were also aware on other associated diseases such as HIV/AIDS that they could encounter as a result of rape. According to them, such information was widely disseminated through trainings that they have attended. While such knowledge was visible, some members also discussed on instances that they have heard on women who did not report to the authorities, for the fear of not trusting the authorities and not wanting the other community/family members knowing their plight. Nevertheless, changes in attitudes were less visible. For example, very few, said that marital violence was a form of SGBV, as they found it as a normal occurrence where, the husband had the right to force himself on his wife if she does not obey him.

Community empowerment programmes has brought in changes in lives of women refugees that has not only influenced in gender equality but also helped in creating new spaces in leadership within families and their community and increased her resilience to survive. Whilst this can be considered as a double burden for women on roles at family and at community levels, this has led women into prominence enabling them to be in positions of power and decision-making in their new roles. This was visible among women who were empowered to become leaders in their communities, while some said that such leadership positions came with costs such as strong opposition by their husbands, male relatives or even other female family members. Yet it gave them an opportunity to work for their communities, talk on their needs and interests and a space in their communities. This type of empowerment was linked to economic empowerment, where women were able to have their own investments or micro-credit opportunities. A shift was also visible in terms of issues such as rape, domestic abuse and gender discrimination that has moved from a 'women's problem to a community problem', where men too were involved in workshops, trainings or discussions. Initially, these were tabooed issues wherein men were not involved or did not want to participate. Today, women find that increasing involvement of men in GBV discussions helped in community awareness and prevention. A participant from a FGD explained the sensitiveness of the topic of rape among men, when they were asked, 'what if their daughters were raped?' According to them, the issue of daughters being raped was something a father was unable to tolerate.

The resilience of the survivor is dependent on the survivor's relationship to the broader society where the pre-requisite is to assess how sexual violence predominates within the women-men relationship. Empowering women through sensitizing on the role of gender issues in the society can increase resilience through the following: first, by highlighting the cultural basis for SGBV and victim blaming, the self-blame component of the survivor can be reduced. Second, knowledge of the sex role socialization assists survivors to change their beliefs and regain their social and personal power and positioning in society influencing positive attitudes as being equal in society. Positive attitudes on the basis of positioning in society greatly impacts resilience, as in the narrative of Idil, where Idil's participation in female support groups assisted her to overcome her misery. According to Idil, being in groups with individuals who had faced similar incidents and who are stigmatized in the community assists in building group solidarity and support among group members. The feeling of overcoming the plight and the associated stigma together, gives courage to the group members. Further, support groups facilitate some survivors to become positive agents for change on women's equality and relieve the feelings of helplessness as a survivor of SGBV. In the case of Idil, she has been able to step forward to gear changes in the community by empowering women survivors of SGBV.

Resilience in displaced contexts, need to focus on the manner in which Somali refugee women survive and are resilient in the SGBV aftermath in their day to day lives. Recognizing that resilience are culture-bound and the invisibility of Somali refugee women in the resilience literature shows the need for an understanding of gender norms that impact their experience of sexual violence and resilience strategies. Resilience of survivors of SGBV is dependent on community support and external interventions. While various measure implemented in the Dadaab Refugee Complex mitigate the occurrence of sexual violence, analysis reveal that these measures have been partially effective (Aubone & Hernandez, 2013). Interventions in refugee camps need to target safety for girls and women throughout their lives in which they conduct their day to day activities. Such change is possible only if changes in women's status will occur at individual, community, cultural, national and international contexts. Changes in attitudes remain a core in SGBV prevention programmes which is undeniably difficult, yet deliberate efforts through campaigns aimed at all members of the country including men and boys can be an effective strategy to prevent violence. Research about cultural influence on sexual violence and girls are evolving. No one checklist or chapter is applicable to address the range sexual violence of violence in multiple settings and varied cultural milieu that facilitate the crime to continue bringing us to the fact that cultures are evolving, so do culturally justified SGBV in refugee settings where culture of impunity is positioned in the hegemonic patriarchal cultures necessitation flexibility in interventions.

Notes

1. Interview with a Refugee Women, Dadaab, 11 December 2013.
2. Interview with a Refugee Women, Dadaab, 9 December 2013.
3. Focused Group Discussion (FGDs) in Dadaab, 13 December 2013.
4. Focused Group Discussion (FGDs) in Dadaab, 13 December 2013.
5. Interview with a Refugee Women, Dadaab, 11 December 2013.
6. Interview with Refugee Women, Dadaab, 14 December 2013.
7. FGD in Dadaab, 13 December 2013

8. FGD, Dadaab, 13 December 2013
9. Interview with Refugee Women, 14 December 2013.
10. FGD, Dadaab, 16 December 2013.
11. Field Notes, December 2013
12. Interview with a Refugee Women, Dadaab, 19 December 2013.
13. Field Notes, February 2014.
14. FGD, Dadaab, 19 December 2013.

References

Abdi, A. M. (2006). Refugees, Gender-Based Violence and Resistance: A Case Study of Somali Women in Kenyan Camps. In E. Tastsoglou & A. Dobrowolsky (Eds.), *Gender, Migration and Citizenship: Making Local, National and Transnational Connections*. Ashgate Publishing Limited.

Aubone, A., & Hernandez, J. (2013). Assessing Refugee Camp Characteristics and the Occurrence of Sexual Violence: A Preliminary Analysis of the Dadaab Camp. *Refugee Survey Quarterly, 32*(4), 22–40.

Badurdeen, F.A. (2020). Resolving Trauma Associated with Sexual and Gender-Based Violence in Transcultural Refugee Contexts in Kenya. In: Crepaz, K., Becker, U., Wacker, E. (Eds.), *Health in Diversity – Diversity in Health*. Springer. Available at: https://doi.org/10.1007/978-3-658-29177-8_12. Accessed July 8, 2022.

Buscher, D. (2017). Refugees, Gender and Livelihoods. In S. B. Zistel & U. Krause (Eds.), *Gender, Violence, Refugees*. Berghahn books.

Cahn, N., & Aolain, F. N. (2009). Hirsch Lecture: Gender, Masculinities, and Transition in Conflicted Societies. *New England Law Review, 44*(1), 1–24.

Canning, V. (2017). *Gendered Harm and Structural Violence in the British Asylum System*. Routledge.

Chakrabarti, S. (2017). *Of Women: In the Twenty-First Century*. Allen Lane.

Chung, K., Hong, E., & Newbold, B. (2013). Resilience among Single Adult Female Refugees in Hamilton, Ontario. *Canada Journal of Refuge, 29*(1), 65–74.

Chynoweth, S. (2017). *Sexual Violence Against Men and Boys: In the Syrian Crisis*. UNHCR.

Cockburn, C. (2004). The Continuum of Violence: A Gender Perspective of War and Peace. In W. Giles & J. Hyndman (Eds.), *Sites of Violence: Gender and Conflict Zones*. University of California Press.

Cockburn, C. (2019). The Continuum of Violence: A Gender Perspective on War and Peace. In R. Jamieson (Ed.), *The Criminology of War* (pp. 357–375). Routledge.

Freedman, J. (2016). *Gender, Violence and Politics in the Democratic Republic of Congo*. Routledge.

Galtung, J. (1971). Structural and Direct Violence: Note on Operationalization. *Journal of Peace Research, 8*(1), 73–76.
Garner, J., & El-Bushra, J. (2004). *Somalia – The Untold Story: The war through the Eyes of Somali Women*. Pluto Press.
Giles, W., & Hyndman, J. (2004). Introduction. In W. Giles & J. Hyndman (Eds.), *Sites of Violence: Gender and Conflict Zones*. University of California Press.
Grieff, S. (2010). No Justice in Justifications: Violence against Women in the Name of Culture. *Religion, and Tradition*, pp. 1–45. Available at: http://humanizm.net.pl/reliviol.pdf. Accessed June 7, 2022.
Islan, N. (2015). *Securing Women's Right to Political Participation through the adoption of Quota System in Somalia*. MA Dissertation, Central European University.
Jensen, M. A. (2019) Gender-Based Violence in Refugee Camps: Understanding and Addressing the Role of Gender in the Experiences of Refugees. *Inquiries Journal, 11*(1). Available: http://www.inquiriesjournal.com/a?id=1757. Accessed July 8, 2022.
Kelly, L. (1993) 'Wars Against Women: Sex Violence, Sex Politics and Militarized State', in Susie Jacobs, S., Jacobson, R. and Marchbank, J. State of Conflict: Gender Violence and Resistance. : Zed Books.
Krause, U. (2015). A Continuum of Violence? Linking Sexual and Gender-based Violence during Conflict, Flight, and Encampment. *Refugee Survey Quarterly, 34*(4), 1–19.
Lenette, C., Borough, M., & Cox, L. (2013). Everyday Resilience: Narratives of Single Refugee Women with Children. *Qualitative Social Work, 12*(5), 637–653.
Menjívar, C., & Perreira, K. M. (2019). Undocumented and Unaccompanied: Children of Migration in the European Union and the United States. *Journal of Ethnic Migration Studies, 45*(2), 197–217. https://doi.org/10.1080/1369183x.2017.1404255
Mwangi, C. W. (2012). *Women Refugees and Sexual Violence in Kakuma Camp, Kenya*. Master Dissertation, International Institute of Social Studies, Hague.
Ozcurumez, S., Akyuz, C., & Bradby, H. (2020). The Conceptualization Problem in Research and Responses to sexual and Gender-Based Violence in Forced Migration. *Journal of Gender Studies, 30*(1), 66–78.
Pulvirenti, M., & Mason, G. (2011). Resilience and Survival: Refugee Women and Violence. *Current Issues in Criminal Justice, 23*(1), 37–52.
Reissman, C. K. (1993). Narrative Analysis. *Qualitative Research Methods Series*. Sage Publications.
Shaheed, F. (2009). Violence Against Women Legitimized by Arguments of "Culture": Thoughts from a Pakistani Perspective. In *Due Diligence and its Application to Protect Women from Violence* (pp. 241–248). Brill. https://doi.org/10.1163/ej.9789004162938.i-300.104

Simon-Butler, A., & Mcsherry, B. (2018). *Defining Sexual and Gender-Based Violence in the Refugee Context*. IRIS Working Paper Series, o. 2/2018. Institute for Research into Superdiversity, Birmingham.

Sokoloff, N., & Dupont, I. (2005). Domestic Violence at the Intersections of Race, Class and Gender: Challenges and Contributions to Understanding Violence Against Marginalized Women in Diverse Communities. *Violence Against Women, 11*(1), 38–64. https://doi.org/10.1177/1077801204271476

Storkey, E. (2015). *Scars Across Humanity: Understanding and Overcoming Violence Against Women*. SPCK Publishing.

UNIFEM. (2005). *Denial, Stigma and Impunity- an Uphill Climb to Stop Gender-Based Violence in Somalia*. Available at: https://reliefweb.int/report/somalia/denial-stigma-and-impunity-uphill-climb-stop-gender-based-violence-somalia. Accessed June 7, 2022.

Walker, M. U. (2009). Gender and Violence in Focus: A Background for Gender Justice in Reparations. In R. Rubio-Marin (Ed.), *The Gender of Reparations: Unsettling Sexual Hierarchies While Redressing Human Rights Violations*. Cambridge University Press.

Wasco, S. (2003). Conceptualizing the Harm Done by Rape: Applications of Trauma Theory to Experiences of Sexual Assault. *Violence, Trauma & Abuse, 4*(4), 311.

Wendt, S., & Zannettino, L. (2014). *Domestic Violence in Diverse Contexts: A Re-examination of Gender*. Routledge.

Index[1]

[1] Note: Page numbers followed by 'n' refer to notes.

B. S. Nayak (ed.), *Political Economy of Gender and Development in Africa*, https://doi.org/10.1007/978-3-031-18829-9

Printed by Printforce, United Kingdom